AN AMERICAN PREMIUM GUIDE TO COIN OPERATED machines

by
JERRY AYLIFFE

DEDICATION

Thanks to Gwen Ayliffe, my best friend, and my wife for her continual encouragement and help. Also, thanks to Dan Alexander, for his support and friendship.

Special Thanks to Becky Lowers
for all black and white
sketches of the Coin Operated Machines

CROWN PUBLISHERS INC.
ISBN 544075

Copyright © 1981 by Jerry Ayliffe, exclusive rights Books Americana Inc., Florence, Alabama 35630. All rights reserved. No part of the book may be used or reproduced in any manner whatsoever without permission, except in the case of brief quotations embodied in critical articles or reviews.

TABLE OF CONTENTS

Acknowledgements...2
Introduction...3
What's So Great About Coin-Operated Collectibles?..............3
I Like 'Em! So Now Where Do I Find 'Em?......................4
How To Avoid Being Ripped Off....................................7
Grading... 10
Repair and Restoration...11
Slot Machines..14
    Slot Machine Index......................................16
    Slot Machine Identification Guide.......................19
Trade Stimulators..93
    Trade Stimulator Index..................................95
    Trade Stimulator Identification Guide...................98
Jukeboxes...158
    Jukebox Index..160
    Jukebox Identification Guide...........................161
Arcade and Amusement Machines...................................190
    Arcade and Amusement Index.............................191
    Arcade and Amusement Identification Guide..............193
Vending Machines..228
    Vending Machine Index..................................229
    Vending Machine Identification Guide...................231
Bibliography and Further Suggested Readings.....................236

ACKNOWLEDGEMENTS

   I would like to thank the following people in the field of coin-operated machines for sharing their knowledge, and allowing me to photograph their machines: Bruce J. Benjamin, Al Breuner, Steve Brooks, Bill Butterfield, Tom Cockriel, Jack and Michael Connolly, Jerry Cordy, Dick Craves, Phil Cunningham, Jim and Susan Davy, Marsh Fey, Steve Gronowski, Larry Johnson, Marshall Larks, Rick Lee, Larry Lubliner, John McWain, Bud Meyer, Allan Pall, Neil Rasmussen, Fredrick Roth, Mr. Russell, Alan Sax. Gary Taplin, Bill and Norene Ward, Wayne and Hilda Warren, Bill and Carol Whelan, and D.R. Williams. These people were all very open and helpful in answering my questions and allowing me to photograph their machines. I would like to single out the following people who extensively shared their expertise. I also appreciate the fact that they allowed me to spread photographic paraphernalia all over their homes and shops to photograph their machines. Thanks again, Steve Brooks, Bill Butterfield, Tom Cockriel, Phil Cunningham, John McWain, and Bill and Carol Whelan. I sincerely hope I haven't left anyone out.

INTRODUCTION

Coin-operated collectibles is one of the newest and most exciting areas of interest in the realm of antiques today. Many states have now legalized older slot machines and coin-operated gambling devices for collecting purposes. Of course, arcade, jukebox, and vending collectibles are legal in all states.

Legalization, coupled with the American people's nostalgia for their past has caused dramatic increases in prices over a relatively short period of time.

The values listed in this book will cover the range in price for a piece in poor, fair, good, and excellent condition. Prices reflect the opinion of the author based on personal buying and selling experience, prices at auction, and advertised prices in various publications. These prices are retail; dealers will of course expect to pay less. This book should be considered to be a guide; prices will vary across the country, and they fluctuate rapidly. Don't despair at the high prices; there are still bargains to be found, and that's one of the things that make this area of interest so much fun.

It may be noted that there are no items listed from foreign countries in this book. This is not to say that these non-American machines have no value. However, the interest in this field is mostly centered around American machines.

As a service to the reader, the author has listed a number of tabloids, periodicals, books, publishers and dealers related to the hobby. This is not intended to be a blanket endorsement of those listings. The intention, rather, is to inform the reader, and give him as much help as possible.

With regard to dates, dates listed are the first year of manufacture, or the first patent date. Realize that some machines didn't reach the market for several years after the first patent date. Also, some machines were changed several times during a production run of many years, and the model shown may not be the first one produced. Where less certainty concerning dates exists, the year is marked "circa".

Good luck in this exciting field of coin-operated Americana.

WHAT'S SO GREAT ABOUT COIN-OPERATED COLLECTIBLES?

When the vending machine business was young, the manufacturers realized that they had to make the machines attractive enough to entice the customers to part with their pocket change. What resulted were some of the most beautiful and intriguing machines our country has ever produced.

Sadly, many of these beautiful machines have been destroyed over the years. Vendors were in the business of selling machines that in turn sold a product, service, or entertainment. In the automobile manufacturing spirit, there resulted the inevitable new model. The new machine was brought on location, and the old machine was often taken to the dump (company orders) and smashed up.

Scarcity of these much sought after beauties from our past is a direct result of this planned obsolescence. People would save old furniture, light fixtures, glassware, and musical instruments because they had a utilitarian value; what possible reason would they have for hanging on to an old coin-operated machine?

Somewhere along the line a small diversified group of people across the country started to recognize the fact that we were throwing away a valuable memory of our past. These beautifully ornate, colorful, whimsical, uniquely throught out pieces of American marketing genius began to have value. Spurred on by the nostalgia craze of the past ten years, these machines have gained a wider and wider field of interest and desirability. Auctions across the country are showing dramatic increases in the prices brought for coin-operated collectibles.

Many investors are buying these prices as an alternative to the stock market, real estate, and other conventional investments. Sotheby Parke Bernet, one of the most prestigious auction houses in the country, has had several very successful antique coin machine auctions.

The numbers of people getting involved in the coin-operated machine hobby are growing everyday. The 1st annual Loose Change Fun Fair held in 1979 is a good example of the interest in this field. This show and sale brought in the neighborhood of 10,000 people, to view an estimated 10 million dollars worth of coin-operated antiques and memorabilia. The 2nd annual Loose Change Fun Fair, held in Pasadena, California was an even bigger success, drawing people from across the country, and even overseas.

Where is all this going? No one really knows; it's safe to say, however, that coin-operated collectibles have found their place in our hearts and their preservation is certain.

I LIKE 'EM! SO NOW WHERE DO I FIND 'EM?

One of the most elusive and difficult things for the collector of coin-operated machines is finding them. Even experienced collectors agree that this is the most challenging thing they face and it gets harder every day.

The first thing to do is start watching the local newspaper classified ads. Some of my best finds have been through this source. Also tabloids devoted to classifieds are sometimes a good source. Recently I purchased a 1929 Mill's slot machine for $925.00 through our local newspaper. Even though the machine needed some clean up and restoration, the retail value is conservatively $2000.00.

Business cards are also very helpful. For a $7.00 to $15.00 investment, the returns can be in the thousands of dollars. It really isn't necessary to be in business to have business cards printed up. The format for the card can be simple; such as the following example:

WANTED

Slot Machines   Vending
Arcade   Advertising
Jukeboxes

BUY-SELL-TRADE
John Smith   966-0955

There is one drawback to business cards; if they lay in a desk drawer, they won't accomplish anything. Pass them out at all possible opportunities. People are interested in these old machines and they will enjoy talking about your hobby.

Several months ago I was in a client's house on a business call. Noticing they had some antiques, I mentioned coin-operated machines. One thing led to another, and they told me about a friend of theirs who had a 3 Jacks machine for sale. I eventually bought the machine for $360.00. With restoration it is now worth over $1200.00.

Publications are good sources of machines, and an especially good way to keep up with prices and trends. One note of caution! Before buying through the mail, read the chaper on how to avoid being ripped off.

"The Antique Trader" is a weekly tabloid with categories on most antiques and collectibles. There is a section on coin-operated machines that also includes listings on books, parts, and repairs. Subscriptions (at this writing) are $17.00 per year, through Antique Trader Weekly, P.O. Box 1050, Dubuque, Iowa, 52001. Their phone number is (319) 588-2073.

An important source of information is "Loose Change Magazine". This is a magazine devoted to the coin machine enthusiast. Articles on all subjects related to coin-operated, including current trends, information, book reviews, and display advertising appear throughout this monthly publication. There is also a special section in the back of the magazine for classified ads. Subscriptions (at this writing) are $25.00 per year through Loose Change, 21176 South Alameda Street, Long Beach, California 90810. Their telephone number is (213) 549-0730.

"Coin Slot Magazine" is another source of articles and advertising devoted to coin-operated devices. Subscriptions at this writing are $25.00 for 12 issues. It is available through the Coin Slot, Bill Harris, P.O. Box 612, Wheatridge, Colorado, 80033.

For the jukebox specialist, "The Jukebox Trader" is a good source of information and classified ads. This monthly publication runs between 6 to 12 pages per copy at a yearly subscription rate of $10.00. Subscriptions are available from Rick Botts, publisher, 2545 S.E. 60th St., Des Moines, Iowa 50317.

A catalogue devoted to automatic musical instruments is published by the Mechanical Music Center. It is well-illustrated, a good price guide, and, best of all, free. For a free copy write to Mechanical Music Center, Inc. Box 88, Darien, Connecticut 06820.

For the person who wants to find coin-operated machines the easy way, The Loose Change Fun Fair has got to be the best bet. In October 1979, and 1980, The Loose Change magazine people sponsored a coin machine extravaganza with dealers from all over the country. While the shows featured all types of antiques, coin-ops were the main focus. This show is a good way to view and purchase from a huge selection. The chances are very good that Loose Change will continue with this annual event.

The magazines and periodicals listed are also a good place to take out ads for specialized wants. For example:

> Wanted: Old arcade:
> Cailoscope, mutoscope
> quartoscope machines, Jim
> Jones, 103 Center Street,
> Arcadia, Wash. (311) 967-0222

Local newspapers and classified ad sheets are good for placing want ads. In this case the ads should be more generalized as follows:

> Wanted: **Old** slot machines,
> gambling devices, penny arcade,
> 78 RPM jukeboxes. (311) 967-0222

Flea Markets and garage sales can be a secret hiding place for great buys. Unfortunately it can often be like looking for the proverbial needle in a haystack. Much depends on attitude; if you are the kind of person who enjoys rummaging at flea markets, this can be a place for some great finds. A good friend of mine goes to the local flea market regularly. Almost monthly he comes up with a good find. On the other hand I always doubt that I will find anything at a flea market; as a result I almost never find good items at fleamarkets.

Flea markets are an excellent place to drop off business cards. People that work these markets are pickers; there is a good chance that they will know of places where old machines are available. Don't be too enthusiastic; enthusiasm can often result in higher prices. Flea market people are born bargainers. There is no need to be insulting or rude, but it never hurts to offer the seller less than his asking price. Antique dealers can be a good source of coin-operated. My car has a problem, in that every time I drive by an antique store, it drives right in. Most dealers run across machines from time to time and this is another good place to leave a card. More and more dealers are beginning to specialize in coin-operated and related items. Check out the yellow pages of your phone book; if there isn't a specialized dealer in your area, there might be one in a nearby large city.

Finding old machines is like many other things; the more effort that is expended, the better the result. All of the sources listed are helpful to one degree or the other. The easier methods of finding machines can also be more expensive, and of course using all of the methods is going to net the greatest return.

HOW TO AVOID BEING RIPPED OFF

The elusive, grand, old coin machine is sitting there in front of you. After searching for weeks, finally that special piece of Americana has turned up. The price is high, but your heart says buy it. This scenario has repeated itself many times for me. It has sometimes resulted in a sinking feeling of remorse and anger. Why? Because I have paid the price for the machine, happily run home to repair it, and as I begin to take it apart, things begin to look wrong. There are parts missing; parts are included that belong on other machines; the piece is not authentic and complete.

Complete, complete, complete in old machines is analogous to location, location, location in real estate. If a machine is incomplete it doesn't mean that it has no value. However, it does decrease the value of the piece. How to avoid this trap? Look carefully at all parts of the machine. Check for empty screw holes that would indicate a part was once attached there. Play the machine several times and look for missing functions. Most importantly, ask the owner of the machine if it is complete. He may honestly not know; however, if he says the machine is complete, and it is not, there is an excellent bargaining position for a lower price.

Obviously some missing or replaced parts are more important than others. A missing paycard on a slot machine is acceptable and reduces the value of the machine very little. On the other hand, a missing coin mechanism on a jukebox would be much more important.

All is not lost if there is a missing part. The publications mentioned in the "How to find 'em" section contain advertisements for various parts available for sale. This can be expensive and time consuming, but it can save the day. Another possibility is recasting a part. In other words, one good part from another similar machine can be used to make a part for your machine. Certain things, such as award cards, reel strips, cabinets, springs and decals, are now being reproduced. The point is, all of these things require time, money, and patience, therefore lowering the value of the machine.

Another extremely important thing to look for is original or authentic condition. Most manufacturers made machines for export. The mechanisms were similar or identical to the American counterpart. This is particularly prevalent in slot machines. It is not unusual to find unscrupulous types who will take an American cabinet and place a much less valuable export mechanism in it. Recently an acquaintance of mine purchased a fairly rare Mills slot machine in excellent condition for a good price of $2700.00. He inspected the mechanism when he was initially looking at the machine it was original and in good condition. Several weeks after taking delivery of the machine he took the back door off to look at the mechanism. He was chagrined to find out that the mechanism had been replaced with an export mechanism. At a recent auction I encountered several machines that were nothing more than a conglomeration of parts made to look like a one of a kind gambling piece. These items were sold for a good price. No doubt a lot of pain and suffering occurred when these new owners found out what they had.

As mentioned earlier, many parts of machines are now reproduced. There are skillful restorations in which only a small part of the machine is original and the rest is reproduced parts. In fact, there are several machines currently on the market that are totally reproduced.

There is a good and bad side concerning reproduced parts. On the positive side, reproduced parts allow restoration of a piece that might otherwise end up as spare parts or, worse, on the junk pile. However, in my opinion, when a machine reaches a point where more than one-third of the parts are reproduced, value declines rapidly.

What's the lesson in all of this? Even experts can make mistakes, and it never hurts to be too careful or too well-informed. It takes time and hard knocks to recognize if something is wrong with a machine. However, I have found that trusting my own inner judgement is as important as everything else. Almost every time I ignore that little inner voice tht says there is something wrong with this machine, I end up with a piece that was a mistake for me.

The preceding advise is based on personal inspection of the merchandise. Mail order buying creates a whole new set of problems and an interesting dilemma.

A great variety of coin-operated machines are offered through "The Antique Trader", "Loose Change", "Coin Slot" and various other publications. Often good machines are offered at very competitive prices. Part of the problem is, not being able to personally inspect the machine. Many sellers will send a photograph, but often that does not tell the whole story either.

Generally the seller wants his money up front and then the buyer waits for the shipping process to receive his merchandise. Here is where an element of patience comes in, because often this process can take months. In one case a seller told me that if I would send him a cashier's check he would send the merchandise to me immediately. After a month of hounding him with expensive long distance phone calls I received one part of the order. After more phone calls I received a refund on the other part of the order. What would have happened if the seller refused to send the merchandise, or refund the money? Well of course there are legal outlets for this, but who wants to get involved with all of that?

However, in all fairness, I don't personally know of any cases where anyone has sent his money and not either received the item or his money back. The point is, sending for expensive antiques through the mail can be a time-consuming gamble. To be completely honest, I must admit that I still buy through the mail. Many of the things I have purchased this way have been very worthwhile. I'm just more cautious than I used to be.

The following are some methods I use to avoid disappointments: When calling on the merchandise, or writing, ask these questions: Is the piece complete? Are there any reproduction or added parts? Is the machine restored inside and out? Is it in good working condition? Are any of the castings or wood parts damaged or flawed? Are the original paycards or information cards intact and in good condition? If I'm disastisfied with the machine will you return my money? At this point negotiations take place and if the buyer is satisfied with the price I suggest two other means of protection: number one, ask the seller to send the merchandise C.O.D. (Both United Parcel Service and the truck lines offer this service). Many sellers will object to C.O.D., but there is a way that both parties can be protected. Offer the seller a deposit equal to the shipping charges so that you both have a vested interest in the sale. Also have the seller specify payment of the C.O.D. order with a certified check or money order. That way the seller knows that the check will be good. Some sellers will go along with this and others will not. In either case I suggest one other means of protection. Have the seller send you a **signed** invoice, with a complete statement of machine condition, and terms of the sale. Another way of accomplishing the same objective is for the buyer to write on the back of his check or money order a statement similar to the following: Endorsement of this check acknowledges full payment for a Watling Blue Seal slot machine with double mint vendor; guaranteed to be complete and original, and in good working order. Failure to send the above machine within 5 working days of receipt of check entitles buyer to a full refund.

It is questionable if the above safeguard would have any weight in a court of law, but it at least will make the seller think twice about trying to rip you off.

All of this may seem paranoid, but sending thousands of dollars through the mail for a machine sight unseen can be very scary. None of the ideas I have listed are failsafe; it is still possible to get burned. However, at least the terms of the sale are weighted a little more in favor of the buyer.

GRADING

Grading is a science within itself. I have overhead 15 minute debates between collectors as to the grade of a particular machine. It's inevitable that once something becomes as valuable as coin-operated machines, there is bound to be very specific guidelines.

I have arbitrarily divided the categories of condition in this book into poor, fair, good, and excellent or restored. No system of grading is perfect. The problem with grading is in trying to be objective about something that also has a subjective side. It is sometimes very hard to fit a specific machine into a category of condition. This is where the subjective, judgemental, fly by the seat of your pants philosophies come into play.

Preceding with the premise that no grading system is perfect, the following is my simplified system. Incomplete means a machine that is missing some of its parts (back doors, coinboxes, handles, reel strips, etc.). The value of this machine would be the same as one in poor condition minus the cost of replacing the missing parts.

Of course an unknown factor comes in as to if the part can be replaced at all. Poor condition means essentially complete, but with broken glass, cracks in castings or weldable parts, and often non-working condition. Fair condition means essentially complete, fair general appearance, working or workable with minor adjustment. Good means a machine which is complete, good appearance inside and out, and in good working condition. Excellent or restored condition represents a machine that has been properly stored for years and has excellent original appearance inside and out. This machine would be complete and in perfect working order. Restored condition represents a machine that has been restored to as new condition, is complete, and operates perfectly.

Another popular grading system published by the **Loose Change** people is the 1-5. One is the best condition, equivalent to excellent or restored and a grade 5 is equivalent to a machine in poor condition. The specifics of this system are available from Mead Publishing in the "Official Loose Change Blue Book". This system is very specific and details the condition required for the various parts of the machine.

It is important to read and understand the various systems of grading. Understanding the language can aid greatly in conversing with other collectors. Grading is also a big help in systematically evaluating a machine. It can sometimes bring out an aspect of a machine, either good or bad, that would have otherwise gone unnoticed.

REPAIR AND RESTORATION

One of the first things people ask when they look at coin-operated machines is "does it work"? It's a valid question indeed because it's very important to the value of a machine. I watched a Rol-A-Top go at auction for $1700 when it's a $3000 machine simply because it stopped working while the auctioneer was demonstrating. Later the very happy buyer found that it was only an easily repairable freeze up created by a bent coin. Unfortunately all problems are not that simple.

One option to repairing machines, particularly slot machines, trade stimulators, gumballs, and arcade machines (which are primarily mechanical) is to search for a person in the area who is able to repair these machines. Unfortunately it is somewhat tough to find these people. If you're lucky enough to live near one of the antique dealers that specializes in coin-operated equipment, they will most likely know of a repair person.

If no such outlet is available, then try an ad in the local newspaper, or the nearest big city newspaper. Sometimes a Mr. Fixit-type person who has a good mechanical aptitude can also be helpful as a last resort.

Jukeboxes and pinballs can usually be fixed by a repairperson who works on the more modern machines. The problem with this, however, is their reluctance to attempt something that they are unfamiliar with. Often a repair manual, which is fairly obtainable in the case of jukeboxes, will be very helpful. I luckily stumbled on to a young electronics student who likes to tinker. He has been able to repair every jukebox and pinball I have run across. The real beauty of this is that he appreciates the work and enjoys doing it. And the really nice thing is that he works at an hourly rate that is not unreasonable. Possibly the best place to find a person like this would be the local college or electronics repair school. Talk with the teacher and ask for a student with a good electronic and mechanical aptitude.

In lieu of having someone else fix the machine, how about fixing it yourself? Going back to my own personal experience, I have always considered my mechanical aptitude to be fairly low. This put me in a bad position in that a number of the machines I run across are in need of repair. The end result is that by necessity I have found that I can fix things I never dreamed possible. I believe this can work for almost anyone. Necessity **IS** the mother of invention. There are some books and manuals that can be helpful in this endeavor. In the repair of slot machines there is a book that I consider to be very important. The title is **The Owner's Pictorial Guide for the Care and Understanding of the Mills Bell Slot Machine.** This book is published by the Mead Company, 21176 South Alameda Street, Long Beach, California, 90810, and it is also available through antique dealers. The price is $24.95 which might seem a little high for a paperback book. However, in my opinion, it's worth every penny. **The Owner's Pictorial Guide** is well-written and illustrated, and my hat goes off to Dan Mead and Robert Geddes for their effort. In my opinion, with a little patience, a fair mechancial aptitude, and this book, a person should be able to take apart and reassemble or repair a Mills slot machine. The Mead Company also has plans to publish similar books on the Watling and Jennings mechanisms.

Many of the original owners manuals for slot machines are available through various publishers and dealers. Coin Slot Books, and The Mead Company both carry owner's manual reprints. Their addresses are listed in the bibliography.

Luckily for jukebox owners many of the owner's manuals for the collectible jukeboxes have also been reprinted. These manuals along with well-illustrated part catalogues are available from Jukebox Junction, Box 1081, Des Moines, Iowa, and Victory Glass Co., P. O. Box 119, Des Moines, IA 50301.

Fortunately for the gumball and vending enthusiast these machines are fairly simple. It's a good thing, because there are very few owner's manuals available on these machines.

Pinball and arcade machine manuals are even harder to come by. So these machines can be very tricky to figure out. However, there is a book for the general repair of pinballs that is listed in the bibliography.

Now that the machine functions, what about its general appearance? A question comes up; should a machine in good original condition be cosmetically restored or not? It basically comes down to what satisfies the owner. The general trend in coin-operated antiques is to restore a machine back to original condition, much like antique automobiles are restored. In the case of machines, this means cleaning and relubricating the mechanism, refinishing any wood parts, and polishing and repainting the metal castings.

The first order of business is to determine if there are any missing parts. If there are, and the owner wants to replace them, I suggest starting the search immediately. Watch for advertisements, or place a "parts wanted ad" in periodicals such as "Loose Change," "The Coin Slot", "Antique Trader", "Jukebox Trader", etc. (Adresses are listed in the bibliography).

The next step in the restoration is to carefully disassemble the machine, although this does not necessarily mean disassembling the entire mechanism. Place screws and bolts with their respective pieces, and carefully label everything.

Pieces that need plating or polishing can now be taken to a reputable plater, preferably one that is used to working with antiques. Dull aluminum pieces can be brought up to a beautiful luster. Nickle plating, if it is pitted should be replated. Don't get talked into replating a nickle piece with chrome; it just isn't the same. Brass pieces can usually be polished back to a high shine. Smaller pieces can be buffed on a home buffing wheel. The motor should be at least ½ horsepower and spin at 3450 RPM's. Buffing wheels and buffing compound can be obtained at rock shops, Sears stores, and sometimes the local hardware. This is a fairly dangerous practice, however, and should be done with caution. Professional platers use a special acid solution to soak articles before buffing. Unfortunately this is impractical and dangerous for home use. For this reason I take my larger buffing jobs to a pro. While the metal pieces are at the plating shop, it's a good time to put the mechanical parts in to soak. Lacquer thinner will work, but carburetor cleaner does a better job. Be sure to follow label directions, use rubber gloves and eye protection, and be careful. Let the parts soak for at least 2 hours; overnight is better. In the case of carburetor cleaner, usually it is washed off. Then, preferably, it should be air dried with a compressor. One advantage of lacquer thinner is that it will dry fairly quickly in the air by itself, thus cutting out the need for an air compressor. Next lubricate moving parts; use light oil, such as sewing machine oil on bearings and smaller moving parts. Vaseline is recommended on larger parts with a heavy abrasion factor. Automotive oils and lubricants are not recommended because they have additives that can cause corrorsion on the plating. Some purists completely disassemble the mechanism and have them bead blasted (a process similar to sandblasting) or blasted with ground walnut shells. Since this takes the plating off, the parts then need to be replated. The machine parts are then reassembled and lubricated as mentioned earlier.

   The next step is to refinish wood parts. First obtain a furniture stripping paste, apply it strictly according to the directions. Again be careful; wear gloves, protective clothing and glasses. Once the piece is stripped and dry, apply a finishing stain such as Watco, Miniwax, etc. Most of these finishing materials are self-sealing, so the next step is a protective finishing material. The preferred material for this is spray finish lacquer. Throw away the brushes, as they will never give the type of final finish that will look professional.

   For those who do not like a high gloss finish, both Watco and Miniwax can be finished with satin wax.

   The final finishing touch, assuming metal parts are back from the plater, is painting. Parts that were painted on the original piece should be repainted with the same color. I personally use spray can lacquer. It dries quickly and is easy to work with. Any overspray that gets into unmasked areas can be cleaned up with lacquer thinner. Be sure to do this in a well-ventilated room, unless you're trying to give yourself permanent brain damage.

   Replace completely worn out or missing pay cards, instruction sheets, reel strips, and decals where possible, with reproduction items. These are supplied by various people who frequently advertise in "Loose Change", "The Coin Slot", "The Antique Trader", "Jukebox Trader", etc. If any of these parts are still intact and fairly legible it is best to save them or put them underneath. They have value and character in themselves. It is always better to save and reuse the original paper pieces and decals where possible.

The final and most rewarding step is to reassemble everything, and admire that gorgeous old machine.

SLOT MACHINES

Slot machines are essentially any machine with a coin slot that involves gambling or speculation. The term "slot machine" is being used because it seems to be the most accepted terminology for coin-operated gambling devices. One arm bandits, bell machines, 3 reelers, uprights, and trade stimulators are all jargon for various machines that fall under the general heading for slot machines. In fact, slot machines in the early days meant anything that was operated with a coin. This section of the book will deal more specifically with automatic payout type gambling devices. This primarily includes the early upright wheel of fortune type machines, and the 3 reelers that are most familiar. There are various other oddities that do not fall into any specific group. These machines which are the automatic payout type, but are not in the upright or three-reeler category are also included in this section.

Slot machines began their predominant history in the 1890's, although there were earlier examples of gambling devices. The most popular machines of that period were the large upright slots with one disk that resembled a wheel of fortune. The player selected a color category or several colors, placed his bet in the appropriate color coded slots, and gave the wheel a spin. If the wheel landed on the selected color, there was an automatic payout.

Automatic payout is an important concept. What was so attractive about automatic payout? First and foremost it gave the storekeeper, saloon owner, restaurant owner or whomever an automatic employee. While the businessman was taking care of business, this machine automatically took the customer's money, entertained him, and paid him if he was a winner in the transaction. This was an employee that was never late for work, was willing to do its job, never needed instruction, and didn't collect an hourly wage.

These upright slot machines were popular with the customer as a source of entertainment that enticed the underlying human interest in risk. The businessman liked them because they were a source of revenue. Slot machine manufacturers did their best to fill that need. In the spirit of bigger and better, there were double and triple models. Some even had music boxes that further entertained the customer while he gambled, and circumvented gambling laws.

In 1905 a San Francisco inventor named Charles Fey revolutionized the slot machine industry. Charles Fey produced the first counter top, 3 reel, payout type slot machine and named it the Liberty Bell. Why did this machine become so popular? For one thing, it was a lot more exciting to watch and to play. There were bigger payoffs and more winning combinations available. The machines were also compact, easy to set on a bar top, a store counter, or a multitude of other locations. The rest is history. Literally millions of machines have been produced using essentially the same format.

Mills was quick to copy the Liberty Bell with their Operator's Bell. Caille and Watling later followed with similar copies and in the 1920's a host of other manufacturers jumped on the bandwagon.

With the advent of all these machines the law and order types felt obliged to stamp them out. Laws were brought about to stop gambling machines, but the manufacturers and operators were not about to give up this highly lucrative venture. What they did instead was to make their machines hide under the guise of being vending machines. They did this by hanging a confection vendor on the side of the machine or incorporating it on the front. Everytime a customer inserted a coin, he had the option of receiving a package of gum or candy. Also if the customer hit a payout, the machine dropped tokens, instead of money. Therefore if the operator wished to exchange the player's token for merchandise, or cash, that was strictly up to him.

Slot machines hit their heyday in the depression. Law endorcement seemed to turn its head and 3 reelers began appearing everywhere. It would seem that during an economic disaster, gambling would be the last thing on people's minds. Human nature, however, turned out to be the reverse. Maybe people felt that gambling was their only chance at beating the system.

During and after this period, the operators became more and more corrupt, and our government took a more active role in enforcing the morality of our country. In 1951 the federal government passed a law prohibiting the transportation of slot machines across state lines. This coupled with stricter enforcement of gambling laws by many states put an end to the 3 reeler's heyday. Nevada and New Jersey now are the only states that allow slot machine gambling.

Where does all of this leave the individual that would like to have a slot machine in his home? Unfortunately, for many years this individual was considered to be a criminal. In 1976 after a great deal of lobbying, the California legislature consented to private ownership of slot machines if the machine was made prior to 1941. Since then, many states have followed suit. It is generally agreed upon by collectors that they don't own their machines for the purpose of gambling. They want these machines for the memory of our past, because they are works of art in their own sense, and they're just plain fun.

SLOT MACHINE INDEX

Illustration page number	Machine name	Price range
19	Buckley-Jennings 1929	$1,100-$2,200
20	Buckley Bones 1936	$2,000-$4,000
21	Caille Detroit Floor Wheel 1898	$6,000-$9,000
22	Caille New Century Detroit Floor Wheel 1901	$8,000-$11,000
23	Caille Floor Roulette 1904	$24,000-$30,000
24	Caille Big Six Special Floor Wheel 1904	$12,000-$15,000
25	Caille Big Six Lone Star Floor Wheel 1904	$24,000-$30,000
26	Caille Centaur Floor Wheel 1907	$7,000-$10,000
27	Caille Superior Jackpot Bell 1928	$1,000-$2,000
28	Caille Silent Sphinx Bell 1932	$1,300-$2,500
29	Caille Cadet Bell 1936	$600-$1,200
30	C & F Baby Grand Bell 1931	$900-$1,800
31	Evans-Mills Conversion Bell c. 1927	$1,200-$2,500
32	Fey Liberty Bell 1905	$26,000-$35,000
33	Groetchen Columbia Bell 1936	$450-$900
34	Jennings Operator's Bell 1920	$1,200-$2,400
35	Jennings Today Vendor Bell 1926	$900-$1,800
36	Jennings Jackpot Bell 1929	$900-$1,800
37	Jennings Victoria Bell c. 1931	$800-$1,600
38	Jennings Little Duke Bell 1932	$1,200-$2,500
39	Jennings Sportsman Golf Ball Vendor Bell c. 1932	$1,000-$2,000
40	Jennings Electrojax 1933 Vendor Bell	$1,200-$2,400 $800-$1,600

Illustration page number	Machine name	Price range
41	Jennings Four Star Chief Bell 1936	$800-$1,600
42	Jennings Ciga-Rola Bell 1937	$600-$1,200
43	Jennings Sportsman Golf Ball Vendor c. 1937	$1,000-$2,000
44	Jennings Dixie Belle Bell 1937	$800-$1,600
45	Jennings Silver Chief Bell 1937	$700-$1,400
46	Jennings Bronze Chief Bell 1941	$700-$1,400
47	Jennings Lucky Chief Bell c. 1945	$600-$1,200
48	Jennings Standard Chief Bell 1946	$600-$1,300
49	Jennings Export Chief Bell 1949	$500-$900
50	Jennings Governor Bell 1964	$500-$1,000
51	Mills Dewey Floor Wheel 1899	$5,000-$8,000
52	Mills 20th Century Floor Wheel 1900	$8,000-$11,000
53	Mills Big Six Floor Wheel 1904	$8,000-$11,000
54	Mills Cricket Floor Machine 1904	$11,000-$14,000
55	Mills Operator's Bell 1910	$4,000-$7,000
56	Mills O.K. Vendor Bell 1922	$1,000-$2,500
57	Mills Front O.K. Bell 1923	$1,100-$2,600
58	Mills Jackpot Bell 1928 (torch)	$1,000-$2,500
59	Mills Baseball Vendor Bell 1929	$1,500-$3,000
60	Mills Jackpot Bell 1929 (pointsettia)	$1,000-$2,500
61	Mills Silent Bell 1931	$1,500-$3,000
62	Mills Silent F.O.K. 1931	$1,000-$2,400
63	Mills Silent Gooseneck Bell 1931	$1,500-$3,000
64	Mills Silent Golden Bell 1932	$1,400-$2,800

Illustration page number	Machine name	Price range
65	Mills Mystery Bell 1933	$800-$2,000
66	Mills Extraordinary Bell 1933	$900-$1,800
67	Mills Silent Gooseneck Skyscraper 1933	$900-$1,800
68	Mills Q.T. Bell (Firebird) 1934	$800-$1,700
69	Mills Q.T. Bell 1935	$800-$1,600
70	Mills Futurity Bell 1936	$1,250-$2,500
71	Mills Bonus Bell 1937	$1,200-$2,400
72	Mills Cherry Bell 1937	$900-$1,800
73	Mills Brown Front Bell 1938	$900-$1,800
74	Mills Vest Pocket 1938	$200-$500
75	Mills Chrome Bell 1939	$700-$1,500
76	Mills Black Cherry Bell 1945	$700-$1,400
77	Mills Golden Nugget Conversion c. 1947	$1,500-$3,000
78	Mills Jewel Bell 1947	$700-$1,400
79	Pace Operator's Bell 1927	$1,000-$2,000
80	Pace Bantam Bell 1928	$800-$1,700
81	Pace-Mills Jackpot Front Conversion Bell 1930	$1,000-$2,200
82	Pace Star Revamp Bell 1933	$900-$1,800
83	Pace All Star Comet-Bell 1936	$700-$1,500
84	Pace Deluxe Chrome Comet Bell 1939	$600-$1,200
85	Pace 8 Star Bell c. 1948	$600-$1,200
86	Rockola Reserve Revamp Bell 1930	$1,100-$2,200
87	Watling Brownie Counter Wheel 1900	$1,750-$3,500
88	Watling Jefferson Counter Wheel 1910	$2,400-$4,750
89	Watling Blue Seal Bell 1929	$900-$1,800
90	Watling Twin Jackpot Bell 1931	$800-$1,600
91	Watling Rol-a-top Bell (coinfront) 1935	$2,000-$4,000
92	Watling Rol-a-top (Bird of Paradise) 1938	$2,200-$4,500

BUCKLEY/JENNINGS CONVERSION BELL
circa 1929

Often small manufacturers, operators, and even some of the larger manufacturers used mechanisms from other manufacturers. They placed their own fronts on other mechanisms, and marketed them under their name. It was simple economics: It was expensive to develop and tool up for a slot mechanism, while it was relatively easy to make a front casting. This particular machine mechanism was produced by Jennings, and the front was added by Buckley.

POOR	FAIR	GOOD	EXCELLENT
$1,100	$1,500	$1,900	$2,200

BUCKLEY BONES
1936

Bones is an amazing machine; it's a countertop payout-type slot machine on which the player actually plays a game of craps. Two sets of spinning disks on the top roll dice. Four successive 7's or 11's would win the player a 100 point gold award. Bones, and Bally's Reliance were actually made by the same company, and they are very similar. Bones, however, are considered to be more rare, and are therefore more desirable.

POOR	FAIR	GOOD	EXCELLENT
$2,000	$2,600	$3,200	$4,000

CAILLE DETROIT FLOOR WHEEL
1898

Caille Detroit is one of the first of the upright one reelers. Detroit featured a 6-way play action which allowed the bettor a number of betting options. Essentially the name of the game was to pick the color the wheel would land on. The player could hedge his bet by picking more than one color.

POOR	FAIR	GOOD	EXCELLENT
$6,000	$7,000	$8,000	$9,000

CAILLE NEW CENTURY DETROIT FLOOR WHEEL
1901

Caille was a big leader in upright machines, and the New Century Detroit is a model that is greatly sought after by collectors. The name New Century came from the fact that the country was heading into the 20th century. New Century was decorated with beautifully ornate castings and quarter sawn oak.

POOR	FAIR	GOOD	EXCELLENT
$8,000	$9,000	$10,000	$11,000

CAILLE FLOOR ROULETTE
1904

Caille Roulette is a highly coveted machine by collectors. The game was unique in that it was a roulette concept as opposed to the upright wheel of fortune theme that was much more common. There were a number of betting combinations similar to the other upright games, except the payout was based on where the roulette ball landed.

POOR	FAIR	GOOD	EXCELLENT
$24,000	$26,000	$28,000	$30,000

CAILLE BIG SIX SPECIAL FLOOR WHEEL
1904

Caille machines have always been popular with collectors because of their beautiful castings and rich cabinets. One quick look at the Big Six Special will bear this out. This rare Special was a variation of the Big Six with larger color spaces. It's interesting to note that almost all of the manufacturers copied each other; Miss and Watling also produced Big Six, with Watling being the originator of the Big Six theme.

POOR	FAIR	GOOD	EXCELLENT
$12,000	$13,000	$14,000	$15,000

CAILLE BIG SIX LONE STAR FLOOR WHEEL
1904

Caille incorporated two of their best machines into one with the Big Six and the Lone Star combination. This model also sported a music box. If the local lawman was concerned about gambling, the operation could just point to the music box, and say the machine gave entertainment for the money inserted in its slots. This beautiful machine featured copper plated castings and a beautiful quarter sawn oak cabinet.

POOR	FAIR	GOOD	EXCELLENT
$24,000	$26,000	$28,000	$30,000

CAILLE CENTAUR FLOOR WHEEL
1907

The Caille Centaur is a beautiful machine with its rich ornate castings that are a Caille trademark. It sported a jackpot which was to later in 3 reeler machines with great popularity. Notice the beautiful, cast claw feet.

POOR	FAIR	GOOD	EXCELLENT
$7,000	$8,000	$9,000	$10,000

CAILLE SUPERIOR JACKPOT-BELL
1928

Caille machines are noted for their classic lines. The Superior can best be described as rich looking; the type of gambling device one would expect to find in a gentleman's club. The Superior Jackpot model is a takeoff from the Superior which featured a revealing female figure on the front. Caille mechanisms were somewhat less reliable than their dependable Mills counterparts.

POOR	**FAIR**	**GOOD**	**EXCELLENT**
$1,000	$1,300	$1,700	$2,000

CAILLE SILENT SPHINX BELL
1932

Caille has always been noted for their classy looking machines. While the Sphinx is not as ornate as earlier machines, it has a certain rich feeling. Painted in its bright colors, it's hard to pass by without inserting a nickel.

POOR	FAIR	GOOD	EXCELLENT
$1,300	$1,700	$2,100	$2,500

CAILLE CADET—BELL
1936

Dynamic styling is evident in the streamlined Cadet, with its unique circular jackpot. The Cadet also featured an escalator that moved from the bottom up and dropped the top coin into the coin tube. This machine was the last of a long line of beautiful and unique gambling devices. Caille Bros. of Detroit went out of business, soon after the Cadet was introduced.

POOR	FAIR	GOOD	EXCELLENT
$600	$800	$1,000	$1,200

C & F BABY GRAND
1931

Miniature machines were popular in the 1930's because they could easily be hidden away when the heat was on. The only machine available that was smaller, was the Mills Vest Pocket, that was on the market at a later date. Amazingly enough, even with the diminutive size, this machine still had a jackpot.

POOR	FAIR	GOOD	EXCELLENT
$900	$1,100	$1,400	$1,800

EVANS/MILLS CONVERSION BELL
circa 1927

Look closely at this machine and the Mills trademarks are evident. This popular cabinet was offered by Fey, Mills, Rockola, and Evans. Evans was a large mail order house in Chicago. This machine is representative of the popular Liberty Bell style.

POOR	FAIR	GOOD	EXCELLENT
$1,200	$1,600	$2,000	$2,500

FEY LIBERTY BELL
1905

The Fey Liberty Bell is a legend, because it was the first 3 reel payout-type slot machine. It was invented by Charles Fey, of San Francisco. Note the symbols are not the familiar bell-fruit type that were later introduced by Mills. Finding one of these is like finding a 3 pound gold nugget.

POOR	FAIR	GOOD	EXCELLENT
$26,000	$29,000	$32,000	$35,000

GROETCHEN-COLUMBIA BELL
1936

The Columbia Bell is a unique little machine. One feature that is particularly interesting is an interchangeable coin setup. It could be easily set up to take penneys, nickels, dimes or quarters. Because of its small size it was great for counters. Of course the small size was also great for hiding it away. Groetchen made a great many varieties of this machine in a long production run.

POOR	FAIR	GOOD	EXCELLENT
$450	$600	$750	$900

JENNINGS OPERATOR BELL
1920

Amongst all of the cast iron and aluminum fronts, Jennings came out with a wood front machine. There's a good possibility that this was done to save money. However, the effect is one of richness and overall beauty. Jennings was a popular manufacturer and a lot of it had to do with their solid, dependable mechanisms.

POOR	FAIR	GOOD	EXCELLENT
$1,200	$1,600	$2,000	$2,400

JENNINGS TODAY VENDOR BELL
1926

Why four columns of mints? More than likely Jennings wanted to make the machine look more like a vending machine than a gambling device. The manufacturer's greatest challenge was to keep their machines under the veil of legality. Without that, their market was severly cut. Note the familiar Jennings Dutch boy cast on the front of the machine.

POOR	FAIR	GOOD	EXCELLENT
$900	$1,200	$1,500	$1,800

JENNINGS JACKPOT BELL
1929

Jennings Jackpot Bell is more commonly referred to as the Dutch Boy. Note the model shown was originally set up with a side vendor. The front paycard very specifically says the machine is a vendor and not a gambling device. Bright blue made the two Dutch boys stand out, creating an eye catching machine.

POOR	**FAIR**	**GOOD**	**EXCELLENT**
$900	$1,200	$1,500	$1,800

JENNINGS VICTORIA-BELL
circa 1931

Jennings Victoria was designed with a conservative touch. Victoria's cabinet styling was also used on the Electrojax, one of the first electrically powered slots. While the Victoria isn't considered to be a classic, it is an attractive, reliable machine.

POOR	FAIR	GOOD	EXCELLENT
$800	$1,000	$1,300	$1,600

JENNINGS' LITTLE DUKE BELL
1932

Jennings' Little Duke was an all-time favorite early in the game with operators, and later, with collectors. It was marketed as a more affordable machine for the operator. Coin denominations were 1 cent, 5 cents, and 10 cents predominantly, which appealed to more conservative players. The unique features of the game were its small size, and the 3 spinning wheels instead of reels.

POOR	FAIR	GOOD	EXCELLENT
$1,200	$1,600	$2,000	$2,500

JENNINGS SPORTSMAN GOLF BALL VENDOR BELL
circa 1932

In the 1930's golf was the popular game of the in crowd. So bless Jennings' little inventive soul, they decided to produce a machine that would appeal to the golf set. This machine was the rage at the country club because it paid off in golf balls instead of money.

POOR	FAIR	GOOD	EXCELLENT
$1,000	$1,300	$1,600	$2,000

JENNINGS ELECTROJAX VENDOR BELL
1933

Jennings introduced the world's first electric slot machine. It seemed like a great step in the progressive modernization of slot machines, but it was not readily accepted by the public. No one knows for sure why; possibly the players just didn't trust it. It came in both table model and floor models, both of which are practically extinct today.

POOR	FAIR	GOOD	EXCELLENT
$1,200	$1,600	$2,000	$2,400

JENNINGS FOUR STAR CHIEF-BELL
1936

The Jennings Chief has beautiful graphics of two Indians on a hunt with pine trees in the background. The Chief Series was the beginning of a long and successful run for Jennings. The Four Star and its predecessor, the One Star, are very similar in design. Both were available in vendor models.

POOR	FAIR	GOOD	EXCELLENT
$800	$1,000	$1,300	$1,600

JENNINGS CIGA-ROLA BELL
1937

Jennings Ciga-Rola is a combination cigarette machine-slot machine. The machine could be operated as a simple cigarette vendor, or a gambler could put his money in and take a chance at winning from 1 to 10 packs of his favorite brand. The machine may not be beautiful, but it certainly is unique.

POOR	FAIR	GOOD	EXCELLENT
$600	$800	$1,000	$1,200

JENNINGS SPORTSMAN GOLF BALL VENDOR
circa 1937

The machine pictured on the right is the 1937 updqte of the earlier Sportsman on the left. The internal mechanism remained much the same. Jennings just streamlined the cabinet and placed the pay card at a jaunty angle.

POOR	**FAIR**	**GOOD**	**EXCELLENT**
$1,000	$1,300	$1,600	$2,000

JENNINGS DIXIE BELLE-BELL
1937

The Dixie Bell was a special order machine that was made for Harold's Club. It was named for Dixie Smith, the wife of the well-known owner of Harold's Club. Quite often larger casinos, even now, have machines with special themes made for their businesses.

POOR	FAIR	GOOD	EXCELLENT
$800	$1,100	$1,300	$1,600

JENNINGS SILVER CHIEF-BELL
1937

The Silver Chief is part of Jennings' long running Chief series. Jennings felt that once they had a winning cabinet theme, they should stick with it. This model came with a chromed front, instead of polished aluminum.

POOR	FAIR	GOOD	EXCELLENT
$700	$900	$1,200	$1,400

JENNINGS BRONZE CHIEF BELL
1941

Jennings was obviously into Indians. This is one of the many "chief" themes produced through the years. One of the highlights of this machine is the bronzed plate at the top, and the 2 large embossed stars.

POOR	FAIR	GOOD	EXCELLENT
$700	$900	$1,200	$1,400

JENNINGS LUCKY CHIEF BELL
circa 1945

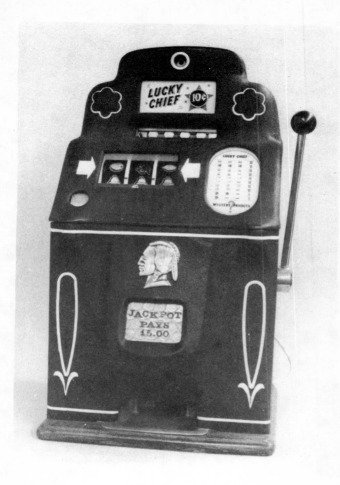

Lucky Chief is another of the series of Chiefs produced by Jennings for many years. The familiar polished casting of the Indian is on the front. The machine had a mystery payout that paid out when the player least expected it, theoretically. The question: Was the Chief lucky for the house, or for the player?

POOR	FAIR	GOOD	EXCELLENT
$600	$800	$1,000	$1,200

JENNINGS STANDARD CHIEF BELL
1946

Jennings Standard Chief is a classic in a sense. While it is not beautiful it is a machine that many old timers recognize as a standard, much like the Mills Hightop. The mechanism is considered to be one of the most reliable ever built.

POOR	FAIR	GOOD	EXCELLENT
$600	$850	$1,050	$1,300

JENNINGS EXPORT CHIEF-BELL
1949

The Export Chief is essentially the same as the Standard Chief. The Export model was set up for foreign coinage, and the Standard Chief was set up for American coinage. Collectors, of course, will pay more for the Standard Chief. Both machines are noted for their reliable mechanisms.

POOR	FAIR	GOOD	EXCELLENT
$500	$650	$750	$900

JENNINGS GOVERNOR BELL
1964

Jennings carried their Indian theme right down to the 1960's. This Governor model has a tic tac toe theme that was first used by Jennings in 1948. Tic tac toe meant that players could win on diagonal as well as horizontal combinations of the tic tac toe symbols. This is an exciting game to play because of all of the possible winning combinations.

POOR	FAIR	GOOD	EXCELLENT
$500	$650	$800	$1,000

MILLS DEWEY FLOOR WHEEL
1899

The Dewey was a successful machine for Mills and was produced for many years. It sported a likeness of Admiral Dewey who was a patriot of the time. Cabinet styles were changed several times over the years.

POOR	FAIR	GOOD	EXCELLENT
$5,000	$6,000	$7,000	$8,000

MILLS 20TH CENTURY FLOOR WHEEL
1900

The end of the 1800's brought on a new age that was our 20th century. Mills knew that the country had high hopes for the 20th century, so they decided it would be a great theme for one of their machines. American marketing has always been a new improved model, and that's what the 20th Century Floor Wheel was about.

POOR	FAIR	GOOD	EXCELLENT
$8,000	$9,000	$10,000	$11,000

MILLS BIG SIX FLOOR WHEEL
1904

Don't be surprised to also see a Watling Big Six, and a Caille Big Six. The original was Watling, but was quickly copied by other manufacturers. The Six theme comes from 6 numbered, and color coded payoff combinations.

POOR	FAIR	GOOD	EXCELLENT
$8,000	$9,000	$10,000	$11,000

MILLS CRICKET FLOOR MACHINE
1904

Mills Cricket is a relatively simple machine based on a principle that resurfaced many times in other gambling machines. The bettor's coin was flipped into a play field of pins and jumped either into a jackpot or a loser's slot. It looked very easy, and was therefore very tempting; however, usually the coin found its way into the last slot. This is an infectious game that people love to play.

POOR	FAIR	GOOD	EXCELLENT
$11,000	$12,000	$13,000	$14,000

MILLS OPERATORS BELL
1910

This cast iron, metal sided Operator's Bell is a direct descendant of the Liberty Bell. Operators were businessmen who purchased machines, convinced other businessmen to allow them to place a machine at their place of business, and then shared in the profits. Mills obviously knew which side their bread was buttered on by making this early machine for the operators.

POOR	FAIR	GOOD	EXCELLENT
$4,000	$5,000	$6,000	$7,000

MILLS O.K. VENDOR-BELL
1922

Generally this machine will be found with a candy vendor on the side. If the vendor isn't there, look for several telltale holes. The O.K. name meant the machine was O.K. for use in non-gambling spots, such as general stores, barber shops, drug stores, etc. The O.K. did this by having the vending feature and a no value type token payout. The little window above the paycard let the player know in advance if the next play was a win or a lose.

POOR	FAIR	GOOD	EXCELLENT
$1,000	$1,500	$2,000	$2,500

MILLS FRONT O.K.-BELL
1923

Slot machine manufacturers were always thinking up ways to get around the gambling laws. The O.K. featured a window that told if the player would win on the next play. Since the player knew he was going to win or lose, the machine was considered to be legal. Notice the 4 columns of mints which allowed the machine to be touted as a vending machine.

POOR	FAIR	GOOD	EXCELLENT
$1,100	$1,600	$2,100	$2,600

MILLS JACKPOT-BELL
1928

Quickly nicknamed The Torchfront because of its torches on either side of the jackpot, this machine was the first of the Mills Jackpot series. Jackpots were features that were included for many years on most machines, because of their great appeal to the clientele.

POOR	FAIR	GOOD	EXCELLENT
$1,000	$1,500	$2,000	$2,500

MILLS BASEBALL VENDOR BELL
1929

Baseball is a truly unique concept in skirting the law and entertaining the customer. The front of this vendor allows for manual scoring of a baseball game. With each spin of the reels, instructions were given for a baseball play. The law was satisfied in that it was a baseball game, not a gambling machine, and the customer was given an additional entertainment feature.

POOR	FAIR	GOOD	EXCELLENT
$1,500	$2,000	$2,500	$3,000

MILLS JACKPOT-BELL
1929

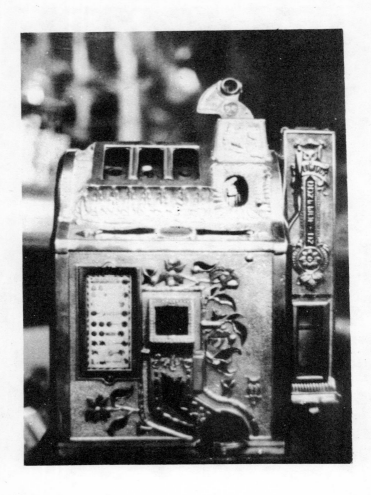

Jackpot got its name obviously because of the jackpot. Sell, was the name of the game with slot machines, and a great inducement to the customer was a visible pot full of coins that could be won. This machine, nicknamed the Poinsettia, was the second model of the Jackpot series.

POOR	FAIR	GOOD	EXCELLENT
$1,000	$1,500	$2,000	$2,500

MILLS SILENT BELL
1931

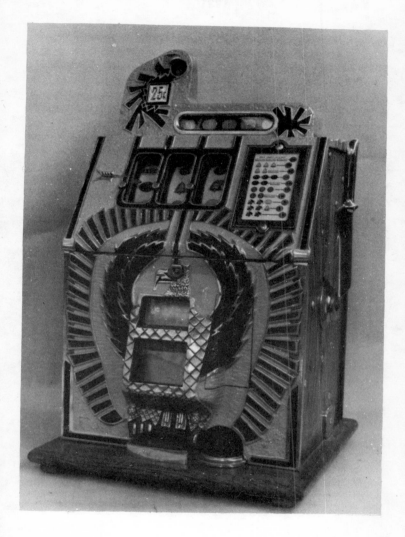

These popular machines are more commonly known as the War Eagle. This was the first of the escalator models with five coins visibly moving along at the top. Mills used the name Silent because it was an improved machine with much quieter operation.

POOR	FAIR	GOOD	EXCELLENT
$1,500	$2,000	$2,500	$3,000

MILLS-SILENT F.O.K.
1931

F.O.K., or Front O.K. meant that the machine had a vendor which made the machine legal, or O.K. In an effort to get around gambling laws, this model was suggested to be a vendor only. The idea was that it vended mints and paid off in trade tokens. Theoretically the vendor portion could not be disconnected or altered.

POOR	FAIR	GOOD	EXCELLENT
$1,000	$1,500	$2,000	$2,400

MILLS SILENT GOOSENECK-BELL
1931

This popular machine has several nicknames, however The Lion's Head or Front is the most common. The Silent Gooseneck was produced along with The Silent. The mechanisms were essentially the same, but the Silent Gooseneck retained the Gooseneck that was prevalent in earlier Mills machines.

POOR	FAIR	GOOD	EXCELLENT
$1,500	$2,000	$2,500	$3,000

MILLS SILENT GOLDEN-BELL
1932

The Mills Silent Golden is more popularly known as the Roman's Head. The Silent Golden name came from the silent mechanism and the gold award feature. Mills marketing genius was next to none and the gold award idea was readily accepted by the public. The idea was to hit 3 gold award symbols, thereby winning the gold award token which exceeded the jackpot in value. The unique graphics on this machine has made it very popular with collectors.

POOR	FAIR	GOOD	EXCELLENT
$1,400	$1,900	$2,400	$2,800

MILLS MYSTERY-BELL
1933

This machine was popularly known as the Blue Front due to its dramatic dark blue background. The name mystery came from the 3-5 payout instead of the earlier 2-4 payout. Winners were mystified when they won 3 coins instead of 2, or 5 instead of 4. Large numbers of this machine were produced, so don't be surprised if one turns up in an unlikely spot.

POOR	FAIR	GOOD	EXCELLENT
$800	$1,200	$1,600	$2,000

MILLS EXTRAORDINARY-BELL
1933

What's so extraordinary about it? It reflected the hope of the new age of modern inventions. The distinctive art deco case was meant to represent technological achievement. It also went under the alias of Gray Front due to its military gray background. Extraordinary was a popular machine that resurfaced as an upright cabinet model in 1938, called the Extraordinary Club Bell.

POOR	FAIR	GOOD	EXCELLENT
$900	$1,200	$1,500	$1,800

MILLS SILENT GOOSENECK SKYSCRAPER
1933

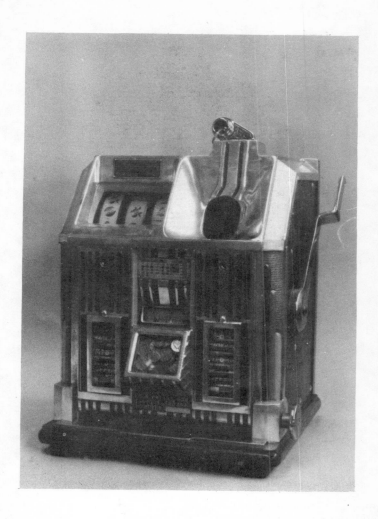

What was Mills doing with a gooseneck in 1933 when they started producing their horizontal coin escalator in 1931? Apparently, someone found a huge quantity of goosenecks laying around doing nothing, and they decided to put their last gooseneck into production.

POOR	FAIR	GOOD	EXCELLENT
$900	$1,200	$1,500	$1,800

MILLS Q.T. BELL (FIREBIRD)
1934

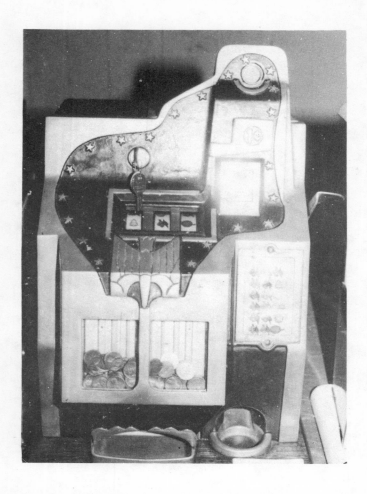

Firebird is the first of the Mills' Q.T. Models. Later they changed to the plain front which was produced in larger numbers. They were usually in 1 cent or 5 cent denominations, and could often be found on display counters, or near cash registers in the 1930's. The psychology of this was much like the gum racks and magazine racks of today. Customers were expected to, and often did part with their extra change at these points of purchase.

POOR	FAIR	GOOD	EXCELLENT
$800	$1,100	$1,400	$1,700

MILLS Q.T.-BELL
1935

The Q.T. was Mills' major entry into the small-sized slot machine market, although they did produce another smaller machine called the Vest Pocket.

Q.T. was an excellent size for slipping under the counter when the heat was on. Remember the expression "on the Q.T." (or on the sly)? It was also useful to operators with limited space requirements. The machine was sometimes coined as the Green Front due to its green color.

POOR	FAIR	GOOD	EXCELLENT
$800	$1,050	$1,300	$1,600

MILLS FUTURITY
1936

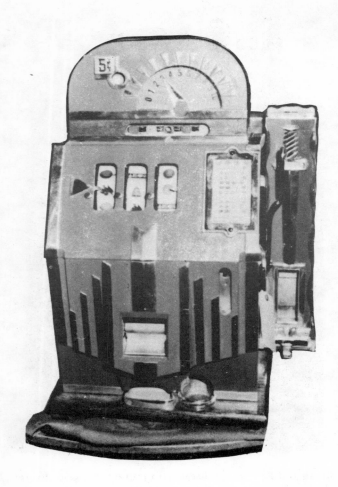

Futurity seemed to be a can't lose machine; the counter advanced on each play that didn't pay out. If the player made 10 passes without any payouts, the machine would give him back all 10 coins. The only catch was getting all the way to 9, and hitting a small payout, because then the counter went back to 0.

POOR	FAIR	GOOD	EXCELLENT
$1,250	$1,650	$2,050	$2,500

MILLS BONUS-BELL
1937

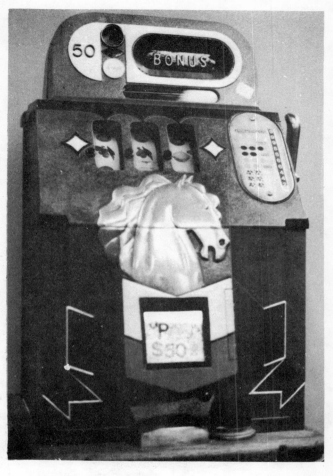

The Bonus is an extremely interesting machine due to the bonus feature. By turning up the letters BONUS in sequence, the player received an 18 coin bonus. The benefit to the house is obvious in that it kept the player coming back for more. The only problem with the machine was its complicated mechanism which had a tendency to fail. This machine was nicknamed the Horsehead for obvious reasons. Note: The machine pictured was most likely an export due to 50 cent denomination; also, Bonus feature is missing.

POOR	FAIR	GOOD	EXCELLENT
$1,200	$1,600	$2,000	$2,400

MILLS CHERRY-BELL
1937

The Cherry got its name from its extra payout of 10 instead of 5 when 2 cherries and a bell or lemon turned up. The Cherry Bell is often confused with the Brown Front which has a similar appearance. Its most common nickname is the Bursting Cherry. Like most of the Mills line, it was produced in a variety of models which included a mint vendor model, and the one shown with an extra jackpot window.

POOR	FAIR	GOOD	EXCELLENT
$900	$1,100	$1,400	$1,800

MILLS BROWN FRONT-BELL
1938

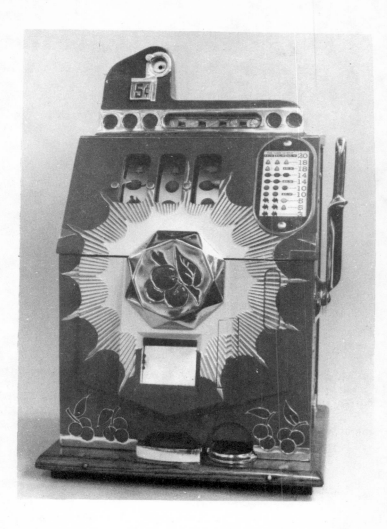

If the Brown Front looks exactly like the Cherry Bell, it's because they are the same cabinet. The main difference between them is that they are painted different colors. The background of the Brown Front is brown, highlighted by a burst of bright orange in the center.

POOR	**FAIR**	**GOOD**	**EXCELLENT**
$900	$1,100	$1,400	$1,800

MILLS VEST POCKET
1938

Vest Pocket was Mills' smallest payout slot machine. It was purposely made boxey and plain to avoid attracting attention. Even the payout card flipped over to conceal the reels. Interesting enough, the mechanism was a micro-version of the larger bell slots.

POOR	FAIR	GOOD	EXCELLENT
$200	$300	$400	$500

MILLS CHROME BELL
1939

Chrome Bell received its name because they came from the factory with a dazzling chrome front, although there were glitter-treated models, and other variations. The nickname that has caught on for this machine is the diamond front, because of the raised diamonds on both sides of the front. This was a successful machine saleswise, and it is considered to be very reliable mechanically.

POOR	FAIR	GOOD	EXCELLENT
$700	$1,000	$1,250	$1,500

MILLS BLACK CHERRY-BELL
1945

Black Cherry is noted for its dramatic raised cherries on the front casting. This machine is sought after by many beginning collectors because it has the look of a classic slot machine. Unfortunately, this machine is defined as illegal in states that have a 1941 cutoff date.

POOR	FAIR	GOOD	EXCELLENT
$700	$1,000	$1,200	$1,400

MILLS-GOLDEN NUGGET CONVERSION
circa 1947

The Golden Nugget Conversion is one of many conversions, or variations of the Mills Jewel Bell. The Golden Nugget Casino wanted their own distinctive slot machine so they converted the cabinet to their own theme. Be careful with conversions; some of the more popular ones have been reproduced. The one pictured above is a **reproduction!**

Original

POOR	FAIR	GOOD	EXCELLENT
$1,500	$2,000	$2,500	$3,000

Reproduction

POOR	FAIR	GOOD	EXCELLENT
$900	$1,100	$1,400	$1,800

MILLS JEWEL-BELL (777)
1947

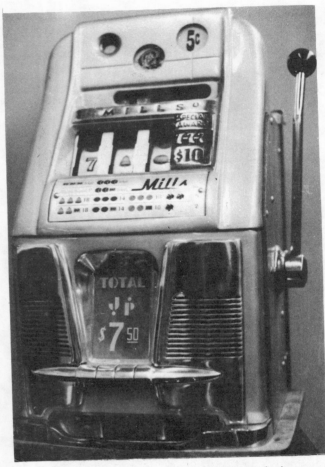

Mills Jewel-Bell is probably the most recognized slot machine ever produced. Because of its durability it is still in operation in many casinos throughout the world. The well-recognized nickname for this machine is the "hightop" due to its tall modular design. Beware of lookalikes; one very common one is the Sega made in Japan. The Hightop was made in a variety of models (the model pictured is the 777) and many of the clubs converted the cabinets to their own themes.

POOR	FAIR	GOOD	EXCELLENT
$700	$1,000	$1,200	$1,400

PACE OPERATOR'S BELL
1927

The Pace Operator's Bell was Pace's first entry into the slot machine business. Pace was a good salesman and his machines grabbed a strong foothold in the market. This machine's unique feature is the circular escalator that Pace continued to use in most of their machines.

POOR	FAIR	GOOD	EXCELLENT
$1,000	$1,300	$1,600	$2,000

PACE BANTAM BELL
(JACKPOT VENDOR MODEL)
1928

Pace's Bantam Bell is the type of cute little machine that little old ladies would like to have setting on top of their Victrolas. It's so cheerful and gay looking, that no one would think badly of it for being involved in gambling. Bantam was sized down to appeal to the small machine market. There were plain fronts, jackpot fronts, and the jackpot vendor front pictured.

POOR	FAIR	GOOD	EXCELLENT
$800	$1,100	$1,400	$1,700

PACE/MILLS JACKPOT FRONT VENDOR CONVERSION-BELL
1930

Castings on this machine were convertible so they could be used by other manufacturers. A telltale sign that the castings were produced by Mills is the 1776 Bell on the gooseneck (which is practically a Mills trademark). The jackpot on this machine is made to look larger than it is by the highly visible graphics surrounding it.

POOR	FAIR	GOOD	EXCELLENT
$1,000	$1,400	$1,800	$2,200

PACE STAR REVAMP BELL
1933

Pace made conversion fronts for operators to change the look of older machines. This particular conversion was made for Mills' machines. The interior mechanism remained the same, so essentially these are Mills machines.

POOR	FAIR	GOOD	EXCELLENT
$900	$1,100	$1,400	$1,800

PACE ALL STAR COMET-BELL
1936

Pace machines are easily recognized by their familiar circular coin escalator. All of the Pace machines featured simple lines and a dependable mechanism. The All Star Comet was produced in a number of variations.

POOR	FAIR	GOOD	EXCELLENT
$700	$1,000	$1,300	$1,500

PACE DELUXE CHROME COMET-BELL
1939

The Pace Comet looks like it could get up and fly. Aerodynamic styling was a sign of things to come. Slot machines became more and more streamlined in design as time went on. The Comet was a successful dependable machine and some are still in operation in various casinos.

POOR	FAIR	GOOD	EXCELLENT
$600	$800	$1,000	$1,200

PACE 8 STAR BELL
circa 1948

Pace machines can still be found in Harrah's Clubs today. Harrah's bought into the Pace organization and produced the 8 Star for use in their own clubs. This machine retains the circular coin escalator that is found on most Pace machines.

POOR	FAIR	GOOD	EXCELLENT
$600	$800	$1,000	$1,200

ROCKOLA RESERVE REVAMP BELL
circa 1930

This machine is a Rockola-Jennings-Fey. The mechanism was produced by Jennings; the conversion front was produced by Rockola, and adapted to the Jennings mech. Charles Fey took the whole unit, ground off the Rockola name on the front, and added his own reel strips. This is an example of how machines and castings were used back and forth by the various companies.

POOR	FAIR	GOOD	EXCELLENT
$1,100	$1,500	$1,900	$2,200

WATLING BROWNIE COUNTER WHEEL
1900

The Brownie, which was produced by both Watling and Mills is a countertop one reeler. It works much the same way as the coveted upright one reelers. Select a color category, drop in a coin, or coins, and if the correct color hits, there was a payout. The machine pictured is missing the all-important top coin acceptor piece.

POOR	FAIR	GOOD	EXCELLENT
$1,750	$2,300	$2,900	$3,500

WATLING JEFFERSON COUNTER WHEEL
1910

Counter Wheels were the smaller counterparts of the upright floor wheels. The concept of the game was essentially the same; they were just made to go on a counter. Watling's Jefferson was a popular machine of the period, and is noted for its nice clean lines.

POOR	FAIR	GOOD	EXCELLENT
$2,400	$3,200	$3,900	$4,750

WATLING-BLUE SEAL-BELL
(JACKPOT FRONT VENDOR)
1929

The Blue Seal comes in a great variety of models: two column vendor, four column vendors, jackpot fronts, plain fronts, etc. It wasn't exactly a classic in terms of beauty, but it worked well, and held up even better. Of course, the Watling company was noted for their fine scales, so this was no great surprise.

POOR	FAIR	GOOD	EXCELLENT
$900	$1,100	$1,400	$1,800

WATLING TWIN JACKPOT BELL
1931

The twin jackpot idea was used by a number of manufacturers. A winner would receive the first jackpot and normally the second jackpot was held in reserve for the next winner. This meant there was always a full jackpot and players could always be assured the possibility of winning a jackpot. This model was part of Watling's Blue Seal line.

POOR	FAIR	GOOD	EXCELLENT
$800	$1,100	$1,400	$1,600

WATLING ROL-A-TOP BELL
(COINFRONT MODEL)
1935

Rol-a-Top, to say the least, is in great demand. The model shown is not particularly rare but collectors love them because of their outstanding graphics and the circular coin escalator. The first model of the long running machine was called the Rol-a-Tor. However, Watling eventually changed the name to Rol-a-Top. The machine has had several different front castings: The horn of plenty spilling out coins (illustrated above), a bird of paradise model with an outpouring of fruit, a mystery front model, similar to the Mills Mystery Bell, a Cherryfront and a rare 5 cent-25 cent model.

POOR	FAIR	GOOD	EXCELLENT
$2,000	$2,600	$3,300	$4,000

ROL-A-TOP BELL (BIRD OF PARADISE)
1938

All Rol-A-Tops are sought after by collectors. Of the many variations that were produced, it would be hard to pick a favorite, but the Bird of Paradise would be a contender. The castings of the Bird of Paradise feature an outpouring of fruit surrounding the Bird of Paradise. All painted in bright colors, it produces a beautiful effect.

POOR	FAIR	GOOD	EXCELLENT
$2,200	$3,000	$3,800	$4,500

TRADE STIMULATORS

Trade stimulators are mechanical games of chance that do not have an automatic money payout. They are based on an endless variety of themes that include spinning wheels of fortune, 5 reel poker games, dice games, roulette games, 21 games, 3 reel symbol games, one reel symbol games, penny drops and the list goes on and on. They come in all shapes and sizes. The early machines were often ornate with beautiful cast iron and rich woods. Later machines made use of cast aluminum and liberal amounts of hardwood were also included on many models.

Trade stimulators began to appear on the scene in fairly large numbers in the 1890's. Basically, stimulators are fairly simple machines. Since they were not encumbered with payout mechanisms, the technology was available fairly early. The early machines were often tied into cigars. For example, the Fairest Wheel of the 1890's was typical of the type of stimulators found on cigar store counters. The customer inserted a nickel in the top, and set a wheel in motion. At the very least, he received a nickel cigar, and if the wheel landed on a 2 or 3, he would receive the equivalent amount of cigars. Who could pass up a deal like that? The store keeper was happy because this device helped to stimulate business, and the parton was happy, because he felt like he got something for nothing.

As time progressed, these machines became popular for bars, pool rooms, hardware stores, and other small businesses. Often trade stimulators were sitting on the counter next to the cash register. This marketing concept is very similar to the merchandise racks at checkout counters in today's stores. Since the customer is in a buying mood, the storekeeper hopes to get a little more of his pocket change.

During the Depression, trade stimulators shared the same popularity as slots. Businessmen had a hard way to go, and anything that helped to stimulate business was welcome. Machines were produced in great varieties during this period by a number of different manufacturers. Popularity began to slow in the 1940's, and by the 1950's, trade stimulators were all but extinct due to their illegality in many states, and lack of interest. One wonders even if they were legal today, if they would have a viable market.

For many years, trade stimulators were passed up by collectors in favor of payout-type slots. However, there is a growing interest in stimulators. Many auctions are showing increased prices and demand for these non-payout machines. Some of the early versions are commanding higher prices than popular 3 reeler payout machines.

The legality of collecting these machines is somewhat unclear. It is fairly well agreed upon that states allowing payout-type machines for collector purposes will allow trade stimulators. The best bet is to check the individual state laws before taking a chance. It's a shame that these unique momentos of our past are misunderstood by certain legislative and enforcement agencies.

This early photograph illustrates two old gents in a smoke shop. There is a Mills Commercial on the counter, and a Mills Jockey against the wall. The barber pole above indicates that a barber shop is in the back room.

In this early bar photograph, the trade stimulator is in the foreground, and is somewhat out of focus. It appears to be a Mills Royal Trader.

TRADE STIMULATOR INDEX

Illustration page number	Machine name	Price range
98	Canda Perfection card machine c. 1900	$500-$1,000
99	Decatur Fairest Wheel Style 2 c. 1900	$500-$1,000
100	Sittman and Pitt card machine c. 1900	$1,200-$2,500
101	Mills Upright Perfection 1901	$600-$1,200
102	The Kelley c. 1904	$1,000-$2,000
103	Mills Pilot c. 1905	$2,500-$5,000
104	Mills Elk c. 1905	$2,000-$4,5000
105	Mills Umpire 1905	$2,000-$4,500
106	Fey On the Level c. 1907	$1,500-$3,500
107	Mills Check Boy c. 1910	$2,000-$4,500
108	Fey 3 Jacks c. 1910	$600-$1,200
109	Mills Puritan c. 1910	$600-$1,200
110	Kitzmiller's Automatic Salesman c. 1920	$250-$500
111	Little Dream Gum Machine c. 1922	$200-$400
112	Caille Fortune Ball Gum Vendor c. 1925	$500-$1,000
113	Mills New Target Practice c. 1925	$350-$650
114	Mills Puritan Bell c. 1925	$400-$700
115	Mills Little Perfection 1926	$400-$800
116	Douglis Puritan Baby Vendor c. 1930	$225-$450
117	Superior Cigarette Ball Gum Vendor 1930	$250-500
118	Fields Blackjack 21 1931	$100-$250
119	Automatic Games The Ace c. 1931	$200-$350
120	Groetchen Dandy Vendor 1932	$250-$550
121	Daval Gum Vendor 1932	$225-$450
122	Exhibit Supply Get-A-Pack c. 1933	$100-$250
123	Pierce Whirlwind 1933 (disc model)	$400-$850
124	Pierce Whirlwind 1933 (reel model)	$350-$650

Illustration page number	Machine name	Price range
125	Pioneer Novelty Big Bones 1933	$200-$400
126	Daval Chicago Club House 1933	$250-$500
127	Rockola Official Sweepstakes 1933	$375-$750
128	Keeney Magic Clock 1933	$225-$450
129	Robbins Baseball c. 1933	$200-$400
130	Groetchen Pok-o-Reel 1934	$225-$450
131	Groetchen High Tension c. 1934	$300-$600
132	Rockola Hold and Draw c. 1934	$250-$500
133	Ft. Wayne Novelty Little Joe Dicer 1934	$200-$400
134	Groetchen 21 Vendor 1934	$275-$550
135	Groetchen Gold Rush c. 1934	$275-$550
136	Groetchen Fortune Teller 1935	$225-$450
137	Groetchen Tavern 1935	$225-$450
138	Buckley Horses 1935	$250-$500
139	Pace New Deal 1935	$250-$500
140	Buckley Cent-a-Pack 1935	$200-$400
141	Keeney Steeplechase c. 1935	$225-$450
142	Garden City Novelty Bar Boy 1936	$200-$350
143	Daval Reel Dice 1936	$100-$200
144	Great States Mfg. Sandy's Horses 1936	$275-$550
145	Garden City Novelty Army 21 Game 1936	$275-$550
146	Daval Reel Spot 1937	$100-$250
147	Garden City Novelty Gem c. 1937	$100-$200
148	Groetchen Punchette 1938	$300-$600
149	Daval Penny Pack 1939	$150-$300
150	Witney Seven Grand c. 1939	$250-$500
151	Jennings Win-a-Pack c. 1939	$75-$150
152	Daval Ace 1940	$85-$175

Illustration page number	Machine name	Price range
153	Daval Buddy c. 1940	$125-$250
154	Groetchen Mercury c. 1940	$85-$175
155	Daval 21 c. 1940	$100-$200
156	Groetchen Yankee c. 1941	$100-$200
157	Dealer's choice c. 1948	$50-$100

CANDA PERFECTION CARD MACHINE
circa 1900

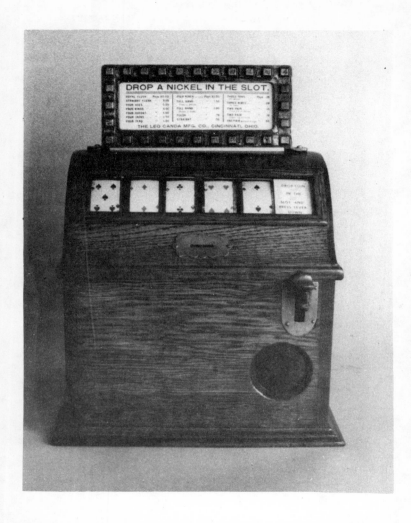

If this machine looks similar to the Mills Perfection it's because they are essentially the same machine. Mills Novelty took over the Canda Company and made this popular machine part of their line. The player was given a poker hand with a spin of the 5 reels, and was paid according to how good the hand was.

POOR	FAIR	GOOD	EXCELLENT
$500	$675	$850	$1,000

DECATUR FAIREST WHEEL
style 2 circa 1900

The Fairest Wheel is a classic cigar trade stimulator. The weight of a nickel placed in the top slot set the wheel moving. Landing on a two, which didn't happen often, gave the customer two cigars. Landing on a one returned the patron one cigar for his nickel.

POOR	FAIR	GOOD	EXCELLENT
$500	$675	$850	$1,000

SITTMAN AND PITT CARD MACHINE
circa 1900

This little beauty had five revolving cylinders with an actual miniature deck of cards attached. The cylinder spun in a random fashion (after the player inserted a coin and pushed the plunger) and a poker hand was shown. With a pair on up, the winner received cigars according to the pay card. The machine was manufactured by Sittman Pitt of New York. However, don't be surprised to see some other jobber or dealer listed on nameplate, paycard, etc.

POOR	FAIR	GOOD	EXCELLENT
$1,200	$1,600	$2,000	$2,500

MILLS UPRIGHT PERFECTION
1901

The Perfection marquee advertised free cigars. For 5 cents the player spun the reels to show a poker hand. The better the hand, the more cigars the patron won. This game could also be played between several customers. Notice the beautiful quarter-sawn oak, and the face cast on the front.

POOR	FAIR	GOOD	EXCELLENT
$600	$800	$1,000	$1,200

THE KELLEY
circa 1904

The Kelley is a super-looking little machine. The front is quarter-sawn oak with a beautiful casting announcing the machine as a Kelley, Chicago Illinois. The object of the game was simple; add up the numbers showing on the reels, and if they equal one of the winning numbers on the paycard, there's a winner. The model shown is the stick gum vendor type.

POOR	FAIR	GOOD	EXCELLENT
$1,000	$1,300	$1,600	$2,000

MILLS PILOT
circa 1905

Mills Pilot is a one reel game that was very similar in concept to the upright floor wheels. The player could bet from one to six coins, and if one of the symbols chosen showed up on the single reel, there was a payoff, usually in trade checks. Notice the beautiful nautical scenes on the castings.

POOR	FAIR	GOOD	EXCELLENT
$2,500	$3,250	$4,000	$5,000

MILLS ELK
circa 1905

Betting from one to five coins varied the odds on this machine at the player's discretion. Winners were usually paid in drinks or cigars. The example shown is missing the marquee and the paycard and symbols are not original.

POOR	FAIR	GOOD	EXCELLENT
$2,000	$2,750	$3,500	$4,500

MILLS UMPIRE
1905

Mills Umpire was based on the ever-popular baseball theme of the day. The player selected a color-coded play for a single, double, etc. If the reel stopped on a strike, out, or foul, it was a loss all the way around. The player had the option of hedging his bet on more than one option, which made the game quite interesting.

POOR	FAIR	GOOD	EXCELLENT
$2,000	$2,850	$3,700	$4,500

FEY ON THE LEVEL
circa 1907

All machines produced by Charles Fey are in great demand, and the On The Level is one of the most sought after. It is made of heavy cast iron and either came painted, or with a nickel finish. The player pushed the plunger, setting the dice in a spin, and then won or lost according to the payout card.

POOR	FAIR	GOOD	EXCELLENT
$1,500	$2,200	$2,900	$3,500

MILLS CHECK BOY
circa 1910

Check Boy is a single reeler counter machine very similar to its larger brother, the Dewey. It had a 6 way play feature by which the player could bet from 1 to 6 coins. Winning was a result of putting a coin in the symbol slot that showed up on the reel. The player was paid in trade checks, thus the name Check Boy seemed logical.

POOR	FAIR	GOOD	EXCELLENT
$2,000	$2,750	$3,500	$4,500

FEY 3 JACKS
circa 1910

This type machine was produced by several different makers: Clauson, Fey, Field, and finally, Rockola. Rockola probably had the largest share of the market with their 3 Jacks, 4 Jacks, and 5 Jacks models. The Clauson and Fey machines are the rarest of the group. Players flipped their coin into a field of pins. The coin danced through the pins to the bottom, and if it ended up in the jackpot slot, there was a winner.

POOR	FAIR	GOOD	EXCELLENT
$600	$800	$1,000	$1,200

MILLS PURITAN
circa 1910

This early trade stimulator resembled a cash register. Was that to make the customer think it was loaded with money, or was it to fool the local lawman? This early model used numbered symbols opposed to the fruit symbol used on the later Puritan Bell. It was also made in cast iron, whereas the Puritan Bell was cast in aluminum.

POOR	FAIR	GOOD	EXCELLENT
$600	$800	$1,000	$1,200

KITZMILLER'S AUTOMATIC SALESMAN
circa 1920

The penny drop machines were produced in great numbers. In the first place, the games were simple; the customer didn't have to be a genius to learn how to play. They were also simple mechanically. There wasn't much that could go wrong, since there were so few moving parts. The player flipped his penny into the playing field, where it bounced through a series of pins. In this particular game, the penny passed through a color-coded target, and was awarded accordingly.

POOR	FAIR	GOOD	EXCELLENT
$250	$325	$400	$500

LITTLE DREAM GUM MACHINE
circa 1922

Little Dream is another of the very popular penny-drop machines. The player received a gumball for every penny. The game was pictured as a baseball game with outs and runs. Of course, the operator could easily convert the run multiples to payout factors. It's also interesting to note that the pennies usually landed in the out slots.

POOR	FAIR	GOOD	EXCELLENT
$200	$270	$340	$400

CAILLE FORTUNE BALL GUM VENDOR
circa 1925

For 1 cent the customer received a gumball and his fortune. The fortune was translated from the pay card according to the bell-fruit symbols that had payoffs similar to the 3 reelers that this machine was based on.

POOR	FAIR	GOOD	EXCELLENT
$500	$650	$800	$1,000

MILLS NEW TARGET PRACTICE
circa 1925

Mills Target Practice was Mills' contribution to the penny drop market. The machine was cast with two discus players on the front; very macho. As with all the other penny drops, the penny bounced through a playing field and into a slot that was good for a payout, or a thank you, maybe next time.

POOR	FAIR	GOOD	EXCELLENT
$350	$450	$550	$650

MILLS PURITAN BELL
circa 1925

Mills Puritan was made for a number of years starting in the early 1900's and running through the 1930's. The one shown is the so-called modern version. It resembled a cash register, possibly to escape the eye of a passing law official. The game itself was a 3 reeler just like slots, except the player was paid across the counter.

POOR	FAIR	GOOD	EXCELLENT
$400	$500	$600	$700

MILLS LITTLE PERFECTION
1926

This Little Perfection is a variation of the 1901 Perfection. The latter model was changed into a flat sided cabinet, similar to many other trade stimulators of the day. Both games were 5 reelers, that spun out varieties of poker hands. Obviously the better the hand, the better the payout.

POOR	FAIR	GOOD	EXCELLENT
$400	$550	$650	$800

DOUGLIS PURITAN BABY VENDOR
circa 1930

A quick look at this machine would lead most people to think it was a Mills, with the Puritan name on the top, and the 1776 liberty bell on the front. Instead, it was produced by Douglis Novelty. There were a number of Baby Vendors on this theme, and it was not unusual for one manufacturer to copy another.

POOR	FAIR	GOOD	EXCELLENT
$225	$300	$375	$450

SUPERIOR CIGARETTE BALL GUM VENDOR
1930

Superior is the typical ball gum vendor, cigarette reel format of the mid-1930's. Interestingly enough, however, it was only a penny play while most others were 1 cent, 5 cent, 10 cent, and even 25 cent plays. For 1 cent, the patron always received a gumball and a shot at free packs of cigarettes.

POOR	FAIR	GOOD	EXCELLENT
$250	$325	$400	$500

FIELDS BLACKJACK 21
1931

Notice on the paycard the very strong warning that this machine is not a gambling device: Believe that, and we've got a bridge to sell you. This was simply a game of flipping a ball bearing in the slots to come up to a total of 21 without going over.

POOR	FAIR	GOOD	EXCELLENT
$100	$150	$200	$250

AUTOMATIC GAMES THE ACE
circa 1931

The Ace is the typical trade stimulator of the period. It is based on a 3 reeler slot machine format. The play was very similar to the one arm bandit except winners were paid over the counter, and there was always a consolation prize of a gumball.

POOR	FAIR	GOOD	EXCELLENT
$200	$250	$300	$350

GROETCHEN DANDY VENDOR
1932

Groetchen's Dandy Vendor is a format that was used by many manufacturers. It includes the slot machine, 3 reeler format, and vends a gumball every time. The player could gamble 1 cent, 5 cents, 10 cents, or 25 cents, and the last coin played was shown in the little window on the right side. These little windows were used to detect slugs, and to let the shopkeeper know how much was gambled in case of a requested payoff. That's right; there was dishonesty even in the 1930's. The gumball viewing area noted strong warnings against using slugs or foreign coins. This particular model comes with a carousel coin divider that ejects every 4th coin to the operator.

POOR	FAIR	GOOD	EXCELLENT
$250	$350	$450	$550

DAVAL GUM VENDOR
1932

Daval is a cute, simple little machine that has a 3 reeler slot machine format. It vended the inevitable gumball as did almost all of the trade stimulators of the time. The reason, of course, was to stay within the scope of the law. It's interesting to note that these gumballs were often close to inedible. Gumballs, of course, were an overhead item, so operators tended to buy the cheapest thing they could find.

POOR	FAIR	GOOD	EXCELLENT
$225	$300	$375	$450

EXHIBIT SUPPLY GET A PACK
circa 1933

Exhibit Supply's Get A Pack was an easy game to play. Two dice were spun and if they showed either a 7 or an 11 the player received one or two packs of cigarettes. A gumball was also vended every time.

POOR	FAIR	GOOD	EXCELLENT
$100	$150	$200	$250

PIERCE WHIRLWIND (DISC MODEL)
1933

Whirlwind, by Pierce, is unique because of its spinning disc instead of reels. Almost all manufacturers of the time used reels, as did the payout slot machines. The disc set up on the Whirlwind is very similar to the Little Duke slot machine.

POOR	FAIR	GOOD	EXCELLENT
$400	$550	$700	$850

PIERCE WHIRLWIND
1933

Whirlwind's art deco case and nice graphic lines make it an attractive machine. One model of the Whirlwind has 3 spinning disks instead of reels, which bears a distinct similarity to the Jennings Little Duke slot machine. The model shown has a visible jackpot that could be released by the operator, for an appropriate winner.

POOR	FAIR	GOOD	EXCELLENT
$350	$450	$550	$650

PIONEER NOVELTY BIG BONES
1933

Big Bones is a unique concept, in that it's a coin-operated chuck-a-luck. A coin trips a release that lets the player turn the glass chuck-a-luck cage. Payoffs are designated on the front paycard. The left upper corner on this example displays a handwritten license number that more than likely was phony.

POOR	FAIR	GOOD	EXCELLENT
$200	$275	$350	$400

DAVAL CHICAGO CLUB HOUSE
1933

Chicago Club House is a reel-type poker game. The 5 reels spun out a game of 5 card poker. Winning hands were shown on the payout card.

POOR	FAIR	GOOD	EXCELLENT
$250	$325	$400	$500

ROCKOLA OFFICIAL SWEEPSTAKES
1933

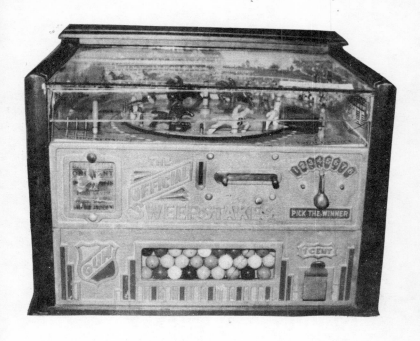

Rockola was thought of as a jukebox company, but they also produced trade stimulators. The player selected a number, and gave the handle a push. Figural painted ponies spun around the track, while a ballbearing spun with them to determine the odds. Selecting the right number was a payoff of the odds times the bet wagered.

POOR
$375

FAIR
$500

GOOD
$625

EXCELLENT
$750

KEENEY MAGIC CLOCK
1933

Magic Clock was an interesting diversion from the typical 3 reel type machine. The player inserted his penny, pulled the lever, and the clock dials spun randomly. If the magic clock pointed at the right symbols, there was a winner or an amused customer.

POOR	FAIR	GOOD	EXCELLENT
$225	$300	$375	$450

ROBBINS BASEBALL
circa 1933

The penny drop theme was used and revised by many manufacturers. Flip a penny into the playing field, and it would bounce through the pins and drop into a slot with or without a point value. This machine was set up with a baseball theme, but there was also an underlying point system that allowed for wagering.

POOR	FAIR	GOOD	EXCELLENT
$200	$100	$300	$400

GROETCHEN POK-O-REEL
1934

Pok-O-Reel is a 5 reel poker hand theme. According to the paycard, various hands are good for free plays. Of course, everyone knew this was for an over the counter payout. Groetchen changed the size and shape of the machine several times over the years, and finished with a fairly small machine.

POOR	FAIR	GOOD	EXCELLENT
$225	$300	$375	$450

GROETCHEN HIGH TENSION
circa 1934

High tension is very similar in cabinet style to its cousin the 21 Vendor. The game theme is horseracing. 3 reels spun, one letter each to spell out a horse's name. If a winning horse turned up, the odds and place finished were shown on the other 2 reels. For the winner it was a payout across the counter and the loser received a gumball.

POOR	FAIR	GOOD	EXCELLENT
$300	$400	$500	$600

ROCKOLA HOLD AND DRAW
circa 1934

Rockola's Hold and Draw advertises 2 spins for one coin. This interesting machine, with its distinctive art deco case is great fun to play. On every spin, the skill buttons can be pressed to stop each reel, thus revealing a poker hand. The player has 2 shots at getting a good hand. Of course, what makes the game fun, is the involvement of pushing the skill buttons.

POOR	FAIR	GOOD	EXCELLENT
$250	$325	$400	$500

FT. WAYNE NOVELTY LITTLE JOE DICER
1934

Little Joe Dicer was a product of the Ft. Wayne Novelty Company. The concept was much the same as the Fey Midget, and other 6 dice spinners. For a 1 cent, 5 cent, or 10 cent bet the player pulled the handle, setting the spinner into motion. The total number showing on the 6 dice was added and compared to the pay card. The hardest combinations, such as 6 of a kind, paid the most.

POOR	FAIR	GOOD	EXCELLENT
$200	$275	$350	$400

GROETCHEN 21 VENDOR
1934

Groetchen 21 is a diversion from the usual 3 reel type machines. It is a five reel 21 game. When the reels are spun, 3 of the reels get covered up leaving 2 in veiw. The player can then hit with up to 3 cards, or stay.

POOR	**FAIR**	**GOOD**	**EXCELLENT**
$275	$375	$475	$550

GROETCHEN GOLD RUSH
circa 1934

Interesting graphics, and the old West theme make Gold Rush an attractive machine. The reel strip combinations corresponded to fortunes to make the machine seem to be a fortune teller instead of a gambling machine. Note the use of one reel with 2 symbols, and the other reel having just 1 symbol.

POOR	FAIR	GOOD	EXCELLENT
$275	$375	$475	$550

GROETCHEN FORTUNE TELLER
1935

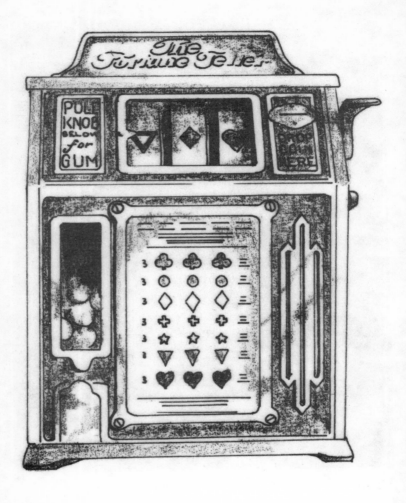

Fortune strips are a familiar con that were used over and over by the gaming industry. The idea, of course, is that the machine only was designed to amuse the customer with a glib fortune printed on the reel strip; and if that wasn't enough, it also vended a gumball. Of course, just for the fun of it, payoffs were shown for certain combinations.

POOR	FAIR	GOOD	EXCELLENT
$225	$300	$375	$450

GROETCHEN TAVERN
1935

Tavern was dedicated to the drinkers of the world. While most trade stimulators were promoting cigarettes or trade checks, the Tavern was putting money in the pockets of the bar owners. The Tavern is the familiar 3 reel concept with the only difference being the beer and whiskey symbols on the reel strips.

POOR	FAIR	GOOD	EXCELLENT
$225	$300	$375	$450

BUCKLEY HORSES
1935

Buckley Horses makes use of 4 reels to simulate gambling at the track. After the player placed his bet, he pulled the handle and gave the reels a spin. If one of the horses on the paycard spelled out on the reels, the winner received the odds amount times the bet.

POOR	FAIR	GOOD	EXCELLENT
$250	$325	$400	$500

PACE NEW DEAL
1935

This gets a little confusing, but both Pace and Pierce made the New Deal. Essentially they are the same machine. The example shown features a jackpot which paid out on 3 deuces. This was not an automatic payout, however; it was controlled by the machine tender. The game theme is 5 card poker, and the paycard on the front of the machines shows the payoffs.

Pace

POOR	FAIR	GOOD	EXCELLENT
$250	$325	$400	$500

Pierce

POOR	FAIR	GOOD	EXCELLENT
$200	$275	$350	$400

BUCKLEY CENT-A-PACK
1935

In the 1890's trade stimulators were used to sell cigars, and 40 years later they were being used to sell cigarettes. The game concept of the Cent-a-Pack was simple, and it showed up again and again. The object was to line up 3 of a kind on the reels for cigarette payouts.

POOR	FAIR	GOOD	EXCELLENT
$200	$275	$350	$400

KEENEY STEEPLECHASE
circa 1935

Steeplechase is an exciting horse race game played with marbles. For a penny, the player can hoist the marbles to the top, and release them from the starting gate. Up to six competitors could play, betting on different colors to win.

POOR	FAIR	GOOD	EXCELLENT
$225	$300	$375	$450

GARDEN CITY NOVELTY BAR BOY
1936

There's no hiding the fact that this machine was made for tavern locations. The first reel on the machine was the odds reel, and the other 3 reels showed beer symbols. Matching 3 of a kind got the player the equivalent number of beers showing on the odds reel.

POOR	FAIR	GOOD	EXCELLENT
$200	$250	$300	$350

DAVAL REEL DICE
1936

Reel Dice was the dice counterpart of the Daval Reel 21. In this game the player spun the reels instead of rolling the dice to make his point.

POOR	FAIR	GOOD	EXCELLENT
$100	$135	$170	$200

GREAT STATES MFG. SANDY'S HORSES
1936

Place your bet on a horse, release the handle, and the horse that finishes in front of the judge's stand is the winner. The paycard says you pay to see it operate, which is a little different than the usual non-gambling disclaimer. Sandy's Horses was a product of Great States Manufacturing.

POOR	FAIR	GOOD	EXCELLENT
$275	$375	$475	$550

Scopitone

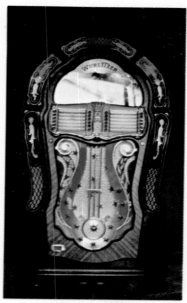

Wurlitzer 1080

Wurlitzer 42   Packard Manhattan

Mills Mystery-Bell

Mills Silent-Gooseneck Bell

Mills Jackpot-Bell

Mills Silent-Fok Bell

Mills Silent-Bell

Mills Black Cherry-Bell

Fey 3 Jacks

Superior Cigarette Ball Gum Vender

Exhibit Supply Get A Pack

Caille Fortune Ball Gum Vender

Mills Upright Perfection

Groetchen Punchette

Jennings Operators Bell

Mills Silent-Golden Bell

Jennings Today-Vender Bell

Watling Rol A Top-Bell

Groetchen-Columbia Bell

Mills Silent Gooseneck Skyscraper

Caille Superior Jackpot Bell

Pace Star Revamp Bell

Caille Silent Sphinx Bell

Pace Bantam-Bell

Buckley Bones

Watling Twin-Jackpot Bell

Pierce Whirlwind

Groetchen Dandy Vender

Daval Gum Vender

Groetchen High Tension

Garden City Novelty-Army 21

Groetchen Pokoreel

Oak Acorn

Northwestern 33

Postage Vender

Abbey Gum Vender

Mansfield Automatic Clerk

Advance Gumball

Wurlitzer 1015

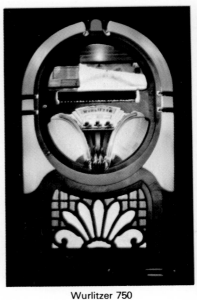

Wurlitzer 750

Wurlitzer 950

Wurlitzer 850

GARDEN CITY NOVELTY ARMY 21 GAME
1936

Army 21 is a great game to play because the player has more to do than just spin the reels. On the first spin, the player's first 2 cards were shown. They then had the option of taking one, or two more cards, by pressing the option buttons. Finally, the dealer's hand button was pressed to reveal the winner. This machine was a product of the Garden City Novelty Company.

POOR	FAIR	GOOD	EXCELLENT
$275	$375	$475	$550

DAVAL REEL SPOT
1937

Daval Reel Spot is a truly unique game that had not been done by anyone previously. That's saying something, because by the 1930's, most of the game themes had been done over and over again. There were four reels; one reel showed odds, and the other 3 reels were covered. After giving the reels a spin, the player guessed which reel would show a spot, and pushed the selection button. If he guessed the right reel, he was paid according to the odds.

POOR	fair	GOOD	EXCELLENT
$100	$150	$200	$250

GARDEN CITY NOVELTY GEM
circa 1937

3 reel cigarette trade stimulators were as common in the 1930's and 1940's as pinball machines are today. Grocery stores, service stations, restaurants, and any other place operators could think of were markets for small trade stimulators. This machine gave a bonus gumball with every spin of the reels.

POOR	FAIR	GOOD	EXCELLENT
$100	$135	$170	$200

GROETCHEN PUNCHETTE
1938

Groetchen played on a radio theme by making their trade stimulator look like a radio receiver. The player inserted a nickel and punched a revolving tape by pulling a handle. The resulting punch was then compared to a radio frequency paycard to determine the payout.

POOR	FAIR	GOOD	EXCELLENT
$300	$400	$500	$600

DAVAL PENNY PACK
1939

Daval's Penny Pack is basically the same machine as the Buddy, which was made again in 1946. Since the cigarette 3 reeler was used over and over again, it without a doubt was very popular with the public. It's not surprising since for 1 cent, the customer had a shot at winning from 1, to 10 packs of cigarettes.

POOR	FAIR	GOOD	EXCELLENT
$150	$200	$250	$300

WITNEY SEVEN GRAND
circa 1939

Witney Seven Grand is a fairly large machine, compared to other trade stimulators of the time. After the bet is placed, and the handle pulled, a bell rings, and the dice are put in motion by the spinning disk they set on. Winners are paid for seven of a kind, down through 4 of a kind.

POOR	FAIR	GOOD	EXCELLENT
$250	$325	$400	$500

JENNINGS WIN A PACK
circa 1939

Jennings Win A Pack will never win a prize for its great beauty. Typical of Jennings, however, the mechanism was smooth and trouble free. The reel strip of the machine illustrated miniature cigarette packs, and the idea of the game was to line up 3 of the same brand.

POOR	FAIR	GOOD	EXCELLENT
$75	$100	$125	$150

DAVAL ACE
1940

Small counter games such as the Ace were seen in large numbers in the 1940's. The concept of this game was a poker hand dealt out by the spinning reels; A straight flush paid the highest. Daval produced a similar machine that had 3 cigarette reels.

POOR	FAIR	GOOD	EXCELLENT
$85	$115	$145	$175

DAVAL BUDDY
circa 1940

The Buddy is representative of the cigarette trade stimulators that were popular in the 1940's. Most everyone would chance a penny for a pack of cigarettes. At the worst, the player was going to get a gumball out of the deal.

POOR	FAIR	GOOD	EXCELLENT
$125	$150	$175	$250

GROETCHEN MERCURY
circa 1940

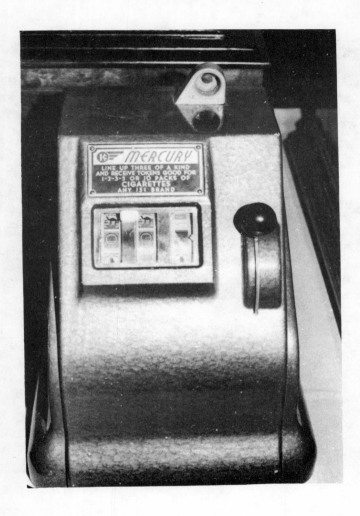

There were thousands of these little trade stimulators throughout the country. It was the popular 3 reel slot machine concept using cigarette label symbols. A player lining up 3 of a kind would receive a token good for from 1 to 10 packs of cigarettes.

POOR	FAIR	GOOD	EXCELLENT
$85	$115	$145	$175

DAVAL 21
circa 1940

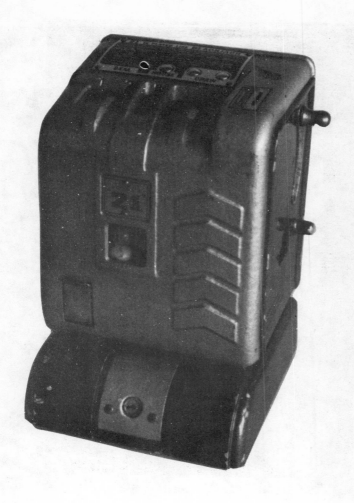

For such a little package, Daval's 21 packed a lot of play. For 1 cent the player spun the reels, and was shown 2 cards. Pushing buttons 3 and 4 would reveal 2 more cards. After this sequence, the House button was pushed and the winner decided.

POOR	FAIR	GOOD	EXCELLENT
$100	$135	$170	$200

GROETCHEN-YANKEE
circa 1941

The Yankee is a cute little 3 reeler with basically the same format as a slot machine. Various combinations on 3 reels gave the player points which could be used in trade with the merchant. Of course, a gumball was delivered every time. Groetchen made similar machines called the Poko Reel, and Klix.

POOR	FAIR	GOOD	EXCELLENT
$100	$135	$170	$200

DEALER'S CHOICE
circa 1948

One of the late trade stimulators made by Associated Distributors of Detroit, this game goes back to a theme used by many early manufacturers, in that it shows a 5 card poker hand. The unique feature of this machine is that it uses actual, full-sized playing cards. With each 5 cent play, the cards shuffle a new card into place.

POOR	FAIR	GOOD	EXCELLENT
$50	$65	$80	$100

JUKEBOXES

Roots for jukeboxes can be found in the coin-operated phonographs of the early 1900's. These coin-ops were nothing more than a phonograph with a coin slot added. At the turn of the century, these devices were a curiosity, and they could be found in arcades throughout the country.

It was really Black American culture that brought the jukebox into vogue. During the 1920's, jazz was beginning to have a groundswell of popularity, especially within the Black community. Conservative White society was dead set against this music, which they considered to be the ranting of heathens. In order to listen to their music, Blacks began to open little juke joints. Since these places were small, and the finances were limited, they were fertile ground for coin-operated record players. Manufacturers and operators began to fill the growing need for coin-operated music machines in the late 1920's, and it was a natural to call them jukeboxes.

Seeburg was one of the first to step into the jukebox market with their 1928 Audiophone. Making use of pneumatics borrowed from their player pianos, Seeburg produced an 8 selection automatic phonograph.

About this same time, Homer Capehart released his automatic phonograph. His phonograph had a larger capacity than Seeburg's, and could play both sides of the record. Capehart was a brilliant businessman, and his future looked bright in the coin-op business. However, business reversals and errors in judgment caused the collapse of his business.

The Wurlitzer family, who had been in the music business for years, began to recognize the potential of the coin-operated phonograph business in the early 1930's. They did two things that would make them the kings of the industry. The first was to hire Homer Capehart, who ran their marketing department, and set up dealerships all over the country. The second, was the later addition of Paul Fuller, who's design leadership in the 1940's produced the most beautiful jukeboxes ever made.

Rockola also joined the growing jukebox market in the 1930's Rockola was actually the owner's last name, not a play on the term rock 'n roll which hadn't even been inspired at that time. Unlike Seeburg and Wurlitzer, Rockola didn't have a background in automated musical instruments. His company produced scales, pinballs, and counter games. However, in the mid-1930's, he recognized the growing market for jukeboxes, and joined the competition.

Other manufacturers jumped on the bandwagon, including Mills, Gabel, AMI, and Filben. In the mid-1930's, the American public recognized the jukebox as an important source of entertainment. Nickels added up to a multimillion dollar industry.

The Golden period of the jukebox was the 1940's. In the spirit of competition, the various manufacturers tried to outdo each other with machines that were more attractive. The result was the most beautiful jukeboxes that were ever produced. Wurlitzer was king in that respect, producing one beautiful model after another. This is the main reason Wurlitzers of the 1940's are the most sought after jukeboxes by collectors.

In the 1950's, the jukebox became the temple of rock 'n roll. Teenagers of this era shaped a new market. The rich woods, and subtle plastics of the 1940's were passe. Their culture commanded brighter chromed boxes, with fins, and outrageous lines. The 78 record was decided to be outdated and cumbersome. It was replaced with the faster, more compact 45's. More selections were required for the wide variety of music that was now available. The jukebox was tied to the hangout, and a general rebellion of adult values.

For the collector this is a sad state of affairs, since jukeboxes of this period have less classic charm. However, jukeboxes of this period are classics in their own right. There is a growing interest in jukeboxes of this era; especially considering the greater availability, and lower cost of these later machines.

In the 1960's the American public began to end their love affair with the jukebox. Many things brought this on; again the rebellious youths of the 1960's rejected traditional values. The Hangout was out, and the sit in was in. A single play on a jukebox was expensive, and many people preferred to invest in an l.p. Superior sound systems became available for home use that outperformed the jukebox as a source of entertainment. Live music regained great popularity in the 1960's. Teenagers preferred listening and dancing to the local 3-piece rock bands that were springing up everywhere.

This trend has continued into the 1970's, and '80's. As a result, the jukebox is becoming extinct. Most people can't remember the last time they played a tune on a jukebox. They are absent from most modern restaurants, and bars, and people don't seem to miss them; or do they:

JUKEBOX INDEX

Illustration page number	Machine name	Price range
161	Capehart 28-G circa 1928	$1,750-$3,500
162	Packard Manhattan 1946	$1,750-$3,500
163	Rockola Counter Model 1939	$500-$1,000
164	Rockola 1422, 1946	$1,250-$2,500
165	Rockola 1428, 1948	$750-$1,500
166	Rockola c. 1955	$375-$750
167	Scopitone 1962	$1,200-$2,500
168	Seeburg Audiphone 1928	$1,750-$3,500
169	Seeburg Audiphone Jr. 1928	$1,750-$3,500
170	Seeburg Symphonola c. 1936	$750-$1,500
171	Seeburg 100B 1950	$375-$750
172	Western Electric Selectraphone 1928	$2,000-$4,000
173	Wurlitzer P12 1935	$750-$1,500
174	Wurlitzer 600 1938	$1,250-$2,500
175	Wurlitzer 24 1938	$1,500-$3,000
176	Wurlitzer 500 1938	$1,250-$2,500
177	Wurlitzer 61 1940	$750-$1,500
178	Wurlitzer 800 1940	$1,750-$3,500
179	Wurlitzer 750 1940	$1,750-$3,500
180	Wurlitzer 71 1941	$1,500-$3,000
181	Wurlitzer 780 1941	$1,750-$3,500
182	Wurlitzer 850 1941	$3,000-$6,000
183	Wurlitzer 950 1942	$6,000-$12,000
184	Wurlitzer 42 1942	$3,250-$6,500
185	Wurlitzer 1015 1946	$2,500-$5,000
186	Wurlitzer 1080 1946	$3,000-$6,000
187	Wurlitzer 1100 1947	$1,200-$2,500
188	Wurlitzer 1800 1956	$375-$750
189	Wurlitzer 1050, 1973	$2,000-$4,000

CAPEHART 28-G
circa 1928

Capehart's 28-G sported an 18 play mechanism. There was no selection; the 18 records played in sequence, and then flipped over to play the other side. Homer Capehart, who founded the company, was later a huge force in the Wurlitzer sales organization.

POOR	FAIR	GOOD	EXCELLENT
$1,750	$2,350	$2,950	$3,500

PACKARD MANHATTAN
1946

Packard Manhattan was created by Homer Capehart. Homer Capehart was a legend in the jukebox industry, working with Rockola, Wurlitzer, and his own Capehart Company at various times. The Manhattan is a huge 20 selection jukebox. Interestingly enough, many of the internal parts are very similar to Wurlitzers.

POOR	FAIR	GOOD	EXCELLENT
$1,750	$2,350	$2,950	$3,500

ROCKOLA COUNTER MODEL
1939

This 12 selection Rockola counter model was made for locations with limited space. There was no internal speaker, so it always needed to be hooked up with a remote system. Later models came with a stand that incorporated the speaker. Since these counter model jukeboxes were so small, internal mechanisms were necessarily quite simple.

POOR	FAIR	GOOD	EXCELLENT
$500	$650	$800	$1,000

MODEL 1422-ROCKOLA
1946

In terms of jukeboxes recognized by the general public as classics, the Rockola 1422 is probably only second to the Wurlitzer 1015. The classic lines of the 1422 were repeated in the 1426 which was brought out the following year. A revolving color cylinder, fully visible record changer, and a 20 selection multiselector were features that also appeared on the Wurlitzer models.

POOR	FAIR	GOOD	EXCELLENT
$1,250	$1,650	$2,050	$2,500

ROCKOLA 1428
1948

This was the last of Rockola's classic, older style boxes. They, like all of the other manufacturers, left the rounded lines, ornate trim, and beautiful woodwork behind. All Rockolas that followed this one had more modern styling.

POOR	FAIR	GOOD	EXCELLENT
$750	$1,000	$1,250	$1,500

1955 ROCKOLA

Is that a 1956 Oldsmobile, or a '55 Rockola? It appears that car manufacturers, and jukebox designers went to the same schools in the 1950's. Fins and chrome were the hallmark of 1950's jukeboxes. Fidelity was greatly improved on this Rockola, as were the selector capabilities.

POOR	FAIR	GOOD	EXCELLENT
$375	$500	$625	$750

SCOPITONE
1962

Scopitone stands by itself as a unique jukebox, and a marketing mistake. For a fairly hefty 25 cents (other jukeboxes of the period were 10 cents for 1 play, or 3 plays for a quarter) the patron got not only the music, but a film of the recording artist synchronized to the music. The films were produced in France, and they were fairly risque for the time. Unfortunately, film production costs were extremely high, and they quickly disappeared from the scene.

POOR	FAIR	GOOD	EXCELLENT
$1,200	$1,600	$2,000	$2,500

SEEBURG AUDIOPHONE
1928

The Regular Audiophone is essentially the same as the Audiophone Jr., except that it has a larger cabinet. Seeburg adapted their model K, E Roll piano cabinet to produce this very early jukebox. Seeburg was always an innovator, and they realized in advance the potential of a coin-operated record player.

POOR	FAIR	GOOD	EXCELLENT
$1,750	$2,350	$2,950	$3,500

SEEBURG AUDIOPHONE JR.
1928

One of the earliest jukeboxes was the Seeburg Audiophone. It had 8 selections operated by pneumatics borrowed from the player piano. The familiar coin slide is there, but it could only swallow one nickel at a time.

POOR	FAIR	GOOD	EXCELLENT
$1,750	$2,350	$2,950	$3,500

SEEBURG SYMPHONOLA
circa 1936

At first glance, this Seeburg looks like an early floor model radio. The selector even looks like a radio dial. The early symphonola mechanism was a 12 selection system that moved the record over in a tray that was picked up by the turntable.

POOR	FAIR	GOOD	EXCELLENT
$750	$1,000	$1,250	$1,500

SEEBURG 100 B
1950

Seeburg was often a leader, as was the case with the 100 B. It was the first 45 rpm - only machine on the market. It followed right behind another first, which was the 100 selection 100A, 78 rpm machine. That was 4 times more choice of selections than the other machines on the market.

POOR	FAIR	GOOD	EXCELLENT
$375	$500	$625	$750

WESTERN ELECTRIC SELECTRAPHONE
1928

Western Electric was owned by Seeburg in 1928, and the Selectraphone is essentially the same as the Seeburg Audiphone. The record selector mechanism was driven by pneumatics, as in player pianos. This jukebox was essentially all mechanical except for the lights, and the amplifier. The Selectraphone would be considered rarer than the Audiphones.

POOR	FAIR	GOOD	EXCELLENT
$2,000	$2,600	$3,200	$4,000

WURLITZER P12
1935

P12 is one of Wurlitzer's earliest entries into the jukebox market. The cabinet is very similar to radio cabinets of the period with a coin mechanism added on. Note the 12 choice simplex selector.

POOR	FAIR	GOOD	EXCELLENT
$750	$1,000	$1,250	$1,500

WURLITZER 600
1938

The model 600 was produced for several years by Wurlitzer and proved a success for them. It employed the use of a rotary selector that was unique to Wurlitzer. It's interesting to note that the record cabinet look started to change at this time into a more flashy, attention-getting machine. This model was also available with keyboard selector.

POOR	FAIR	GOOD	EXCELLENT
$1,250	$1,650	$2,050	$2,500

WURLITZER 24
1938

The model number was significant in that it indicated the number of selections. This was Wurlitzer's first 24 selection machine with their rotary multi-selector. These machines are in short supply because many were dismantled to make rotary Victories.

POOR	FAIR	GOOD	EXCELLENT
$1,500	$2,000	$2,500	$3,000

WURLITZER MODEL 500
circa 1938

The model 500 was a successful jukebox for Wurlitzer, and they continued production on it for several years. While not as showy as later models, it has the classic Wurlitzer charm. Notice the more liberal use of colorful plastics; that was a sign of things to come.

POOR	FAIR	GOOD	EXCELLENT
$1,250	$1,650	$2,050	$2,500

MODEL 61-WURLITZER
1940

The streamlined compact model 61 was one of Wurlitzer's highest selling counter models. It had 12 selections, as did all of the Wurlitzer counter models. Although in a small package, it featured the same beautiful woods, and excellent design, as its larger counterparts.

POOR	FAIR	GOOD	EXCELLENT
$750	$950	$1,250	$1,500

MODEL 800 WURLITZER
1940

In the year 1940, Wurlitzer was prolific in its production of jukeboxes. They produced eleven new models under the excellent design leadership of Paul Fuller. Over 11,000 of the model 800 were produced. The inside mechanism for the Wurlitzers was essentially the same. All models featured a 20 selection multiselector, and a visible turntable system. The model shown is in unrestored condition.

POOR	FAIR	GOOD	EXCELLENT
$1,750	$2,350	$2,950	$3,500

WURLITZER 750
1940

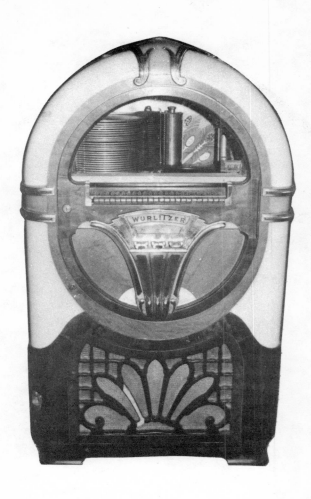

Another of Paul Fuller's great design classics. It's hard to believe that one designer could have produced so many great jukeboxes in one year. The 750 was just one of the many different classics that rolled out of Wurlitzer's factory in 1940.

POOR	FAIR	GOOD	EXCELLENT
$1,750	$2,350	$2,950	$3,500

COUNTER MODEL 71-WURLITZER
1941

Wurlitzer's advertising campaign suggested such obscure locations as barbecue stands, excursion boats, and filling stations as good prospects for a counter model. The model pictured shows the matching stand with its beautiful wood, and graphics work.

POOR	FAIR	GOOD	EXCELLENT
$1,500	$2,000	$2,500	$3,000
		Stand	
$500	$1,000	$1,500	$2,000

780 WURLITZER
1941

The 780 is better known as The Colonial. It was designed for the location that wanted a more conservative appeal. The Wurlitzer sales literature described it as a masterpiece, rich in old world charm.

POOR	FAIR	GOOD	EXCELLENT
$1,750	$2,350	$2,950	$3,500

MODEL 850 WURLITZER
1941

Better known as The Peacock, the Wurlitzer 850 was a Paul Fuller-designed classic. Fuller was prolific in his creation of unique and beautiful Wurlitzers. Approximately 10,500 of these machines were manufactured and they were eagerly sought after by operators. The Peacock is still eagerly sought after by collectors all over the world.

POOR	FAIR	GOOD	EXCELLENT
$3,000	$4,000	$5,000	$6,000

WURLITZER MODEL 950
1942

King of Kings is the only way to describe the Wurlitzer 950. Two things make this jukebox the most coveted by collectors in general. One is rarity, due to the fact that less than 3500 were delivered from the factory. The second reason is the beautiful overall appearance.

POOR	FAIR	GOOD	EXCELLENT
$6,000	$8,000	$10,000	$12,000

WURLITZER MODEL 42 VICTORY
1942

If Wurlitzer's model 950 is the king of collectible jukeboxes, the Victory has to be the queen. This machine was put together with more or less spare parts since raw materials were being used for the war effort. Rarity and beauty equal desirability and that sums it up for the Victory.

POOR	FAIR	GOOD	EXCELLENT
$3,250	$4,250	$5,250	$6,500

WURLITZER 1015
1946

Wurlitzer 1015 is probably the classic of classic jukeboxes. When Hollywood wants a showy jukebox, they usually call on the 1015's for their models. At the end of the War, Wurlitzer pulled out all of the stops, and told Paul Fuller to give it his best shot. The result was the 1015, and it was the most successful jukebox of all time.

POOR	FAIR	GOOD	EXCELLENT
$2,500	$3,250	$4,000	$5,000

MODEL 1080 WURLITZER
1946

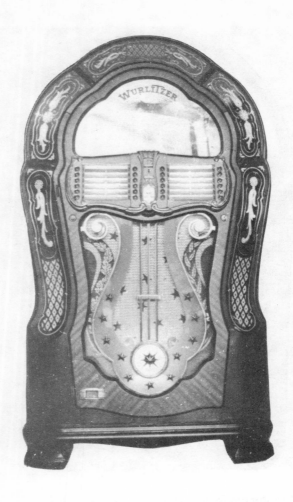

Paul Fuller designed the 1080 with more conservative establishments in mind. Beautifully mirrored graphics replaced the color wheels and bubbler tubes of the 1015. 1015's outsold the 1080, seven to one.

POOR	FAIR	GOOD	EXCELLENT
$3,000	$4,000	$5,000	$6,000

MODEL 1100-WURLITZER
1947

The Space Age design of the 1100 was a sign of the times. Post war America had the dream of a modern, streamlined new era. The 1100 was the last Wurlitzer produced under the design leadership of Paul Fuller. Bubbler tubes, fancy metal castings, richly colored plastics, and beautiful wood inlays were sadly absent from this model. Although the 1100 is a beautiful machine in its own right, it sadly marked the end of an era.

POOR
$1,200

FAIR
$1,600

GOOD
$2,000

EXCELLENT
$2,500

WURLITZER 1800
circa 1956

It's interesting that when most people think back to the Rock and Roll era, they think of classic jukeboxes like the 1015.

Actually the classic jukeboxes were pre-rock and roll. More modern looking machines with streamlined lines characterized the Rock and Roll era. The 1800 is not a classic in terms of styling, yet it is an excellent representative of the heyday of rock and roll. One hundred and four record selections were available and that pleased the operators and the patrons.

POOR	FAIR	GOOD	EXCELLENT
$375	$500	$625	$750

WURLITZER 1050
1973

The 1050 was Wurlitzer's last ditch effort before going out of the jukebox business. It was the company's hope that a nostalgia-evoking machine would help pump resources back into the company's treasury. Unfortunately, the 1050 was not enough, and the company's jukebox division finally gave up. Like many artists who never live to see their work appreciated, the 1050 is now in great demand by collectors.

POOR	FAIR	GOOD	EXCELLENT
$2,000	$2,600	$3,200	$4,000

ARCADE AND AMUSEMENT MACHINES

Arcade and amusement equipment, like most of the other coin-operated machines, began to appear in significant numbers in the 1890's. Centers called arcades were started, that housed a number of amusement machines under one roof. During this early period they were called penny arcades because everything was operated by a penny. These arcades experienced great success because Americans were apparently hungry for entertainment.

Arcades quickly began to fill up with a number of entertainment devices. Edison's coin-operated phonograph was a favorite. For 1 cent the customer could listen to a prerecorded musical performance. Drop card machines showed exotic, faraway places, disasters, and erotic peeks at naughty ladies. Flip card machines, a product of Mutoscope, gave the illusion of motion. Lung testers tested the patrons' blowing capacity, and registered it on a big dial. Strength testers used spring tension to register a customer's grip strength. Illusion machines, and fortune tellers gently conned the player by offering him mystical information. Electrical shock treatments were sold with various inducements; some offered medical cures for multiple ailments, while others were endurance tests. Iron claws were all time favorites. The player tried to maneuver a crane with a hinged scoop into place, scoop up a prize, and deposit it in the reward chute. Often the bucket had a tendency to spill before making it to the reward chute. In spite of the pitfalls, people still kept coming back for more. Orchestrians were also popular, offering the entertainment of what sounded like a marching band.

Arcades were not the only places where amusement games and arcade equipment could be found. They were also present in saloons, poolrooms, barber shops, and other gathering places. These amusement machines were the common American's source of entertainment.

During the 1930's, pinball games started to become popular. The term pinball came from the metal posts, or pins, that the ball dodged through. Early pin games were much smaller than their modern counterparts, and they did not have flippers. The only real control the player had was in how hard he fired the ball. The scoring holes were numbered, and by a manual count, the total score was determined.

Pinball games are significant for one important reason: Pinballs are the only amusement game whose popularity has endured into the present time. Many arcade owners didn't include pinballs in their inventory because they wanted to maintain a family atmosphere. As things progressed in the 1930's, however, it was mostly the so-called less desirable types that were frequenting arcades.

Arcades began to disappear in the late 1930's. The ones that were left in the 1940's and 1950's were mostly connected with amusement parks. The motion picture show had taken over as the entertainment medium of the masses. Eventually, most of the classic amusement games lost popularity to the pinballs.

The interest in pinball games is at an all-time high today. Arcades have resurfaced with electronic marvels made possible through silicone chips. While pinballs are still as popular as ever, they are being joined by a host of video games whose game themes have endless possibilities. Although the games and their technology have changed, playful Americans are still willing to part with their pocket change for fun.

ARCADE AND AMUSEMENT MACHINES INDEX

Illustration page number	Machine name	Price range
193	Clam Shell Mutoscope c. 1895	$700-$1,600
194	27" Regina Upright Style 8A Changer 1896	$7,500-$15,000
195	Mills Punching Bag c. 1900	$750-$1,500
196	Mills Genuine Illusion Machine c. 1900	$2,500-$4,000
197	Caille Peep Show c. 1900	$600-$1,200
198	Clam Shell Mutoscope c. 1900 (Indian front)	$1,000-$2,000
199	Electricity is Life c. 1902	$900-$1,800
200	Mills Madame Palmist c. 1904	$2,000-$4,000
201	Cremona Style K Orchestrian c. 1908	$16,000-$25,000
202	Mutoscope Punch-a-Bag c. 1910	$900-$1,500
203	Uncle Sam Grip Test c. 1910	$2,000-$4,000
204	Mills Violano Virtuoso c. 1912	$9,000-$15,000
205	Mills Wizard Fortune Teller c. 1920	$300-$600
206	Future Products Fortune Teller c. 1920	$200-$500
207	Exhibit Supply Cupid's Post Office c. 1920	$1,000-$2,000

Illustration page number	Machine name	Price range
208	Exhibit Supply Grandfather's Clock c. 1927	$900-$1,800
209	Fey Shoot the Duck 1928	rare
210	Chester Pollard Play the Derby c. 1930	$1,500-$3,000
211	Exhibit Supply Striking Clock c. 1930	$600-$1,200
212	Jennings Comet c. 1930	$350-$700
213	Seeburg Chicken Sam c. 1931	$450-$900
214	Bingo 1931	$250-$500
215	Gottlieb Whizz Bang 1932	$250-$500
216	Gottlieb Big Broadcast 1933	$250-$500
217	Radio Station c. 1933	$250-$500
218	Cupid's Post Office c. 1935	
219	Babe Ruth Baseball c. 1935	$300-$600
220	Challenger Duckshoot c. 1940	$100-$200
221	Exhibit Supply Novelty Merchantman c. 1942	$600-$1,200
222	Mills Owl 1942	$300-$600
223	Mutoscope It's a Knockout c. 1949	$500-$1,000
224	Booz Barometer c. 1950	$75-$150
225	Keeney's League Leader 1951	$250-$500
226	Gottlieb 4 Belles 1954	$250-$500
227	Gottlieb Gold Star 1954	$200-$400

CLAM SHELL MUTOSCOPE
circa 1895

The Mutoscope gave the viewer the closest thing to a moving picture available at that time. A series of pictures showing different positions in a scene flipped in sequence, giving the illusion of motion. There were comedies, disasters, and so-called art studies.

POOR	FAIR	GOOD	EXCELLENT
$700	$1,000	$1,300	$1,600

27" REGINA UPRIGHT STYLE 8A CHANGER
1896

This beautiful music box was a forerunner to the jukebox. The selector knob allowed for 12 different selections of the huge metal disks that were lifted into the upper cabinet by the changer. Two ornately carved dragons stood guard over the glass viewing window.

POOR	FAIR	GOOD	EXCELLENT
$7,500	$10,000	$12,500	$15,000

MILLS PUNCHING BAG
circa 1900

Somewhere in the late 1940's, punching bags began to disappear. Many overzealous customers were missing the bag, and slugging the hardwood, or the metal frame. A lot of broken hands were occurring, and lawsuits were beginning to get popular. Interestingly, metered punching games are beginning to reappear, especially in Western bars.

POOR	FAIR	GOOD	EXCELLENT
$750	$1,000	$1,250	$1,500

MILLS GENUINE ILLUSION MACHINE
circa 1900

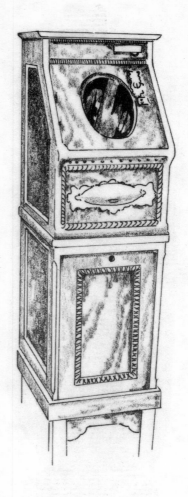

This rare machine was a clever way of tempting a passerby's curiosity. The inducement was to see yourself in the future. After inserting a coin, and pushing the plunger, a 3 dimensional image appeared of a skeleton. There were other themes, and they were all basically cons. The marks knew they were cons, except curiosity got the better of them, and the coin boxes filled up quickly.

POOR	FAIR	GOOD	EXCELLENT
$2,500	$3,000	$3,500	$4,000

CAILLE PEEP SHOW
circa 1900

This early peep show is very unique. For one penny, the shutter on the front of the house raised, and the viewer was treated to a 3 dimensional view of a lady preparing for bed.

POOR	FAIR	GOOD	EXCELLENT
$600	$800	$1,000	$1,200

CLAM SHELL MUTOSCOPE INDIAN FRONT
circa 1900

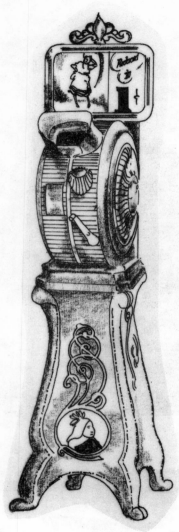

Mutoscrope Indian Front is essentially the same machine as the Clam Shell. The main difference is an Eagle cast into the side, and an Indian on the front. Why the Indian front? Possibly Mutoscope was capitilizing on the country's interest in earlier Americana.

POOR	FAIR	GOOD	EXCELLENT
$1,000	$1,300	$1,600	$2,000

ELECTRICITY IS LIFE
circa 1902

Why would an electric shock treatment be popular at an arcade? In the early part of the century, electricity held a certain magic quality. Uses for electricity were still being discovered. Also note the challenge on the front of the machine: Who can hold the most?

POOR	FAIR	GOOD	EXCELLENT
$900	$1,200	$1,500	$1,800

MILLS MADAME PALMIST
circa 1904

Deposit your money, and ask Madame a question. Pull the appropriate man, or woman lever, and her eyes move from side to side, and a fortune card pops out of the slot. According to the original sales literature, the fortunes were prepared by a noted palmist.

POOR	FAIR	GOOD	EXCELLENT
$2,000	$2,700	$3,400	$4,000

CREMONA STYLE K ORCHESTRIAN
circa 1908

This beautiful upright orchestrian featured eight instruments. For 5 cents the client was entertained with the next tune on the roll. As the holes on the roll passed by a tube, a pneumatic was collapsed, and a corresponding instrument played. What an ingenious way to make music.

POOR	FAIR	GOOD	EXCELLENT
$16,000	$19,000	$22,000	$25,000

MUTOSCOPE PUNCH A BAG
circa 1910

What a great way to get rid of aggression and help to pay some operator's bills. After inserting a coin, the bag was pulled back towards the customer. The puncher then let go with his best shot which registered on the Punch a Bag meter.

POOR	FAIR	GOOD	EXCELLENT
$900	$1,100	$1,300	$1,500

UNCLE SAM GRIP TEST
circa 1910-1978

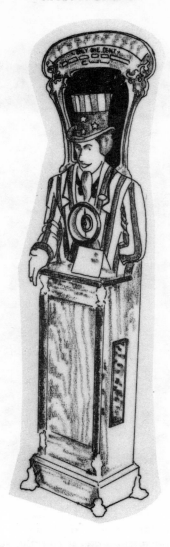

Uncle Sam Grip Test is an all-time favorite arcade machine. For 1 cent the patriot could test the strength of his grip. Beware: The original has been remade several times over the years.

Original

POOR	FAIR	GOOD	EXCELLENT
$2,000	$2,600	$3,200	$4,000

MILLS VIOLANO VIRTUOSO
circa 1912

This amazing musical entertainment machine was a combination violin and piano concert. A perforated paper roll allowed wires on either side to make contact at the appropriate time and perform the functions of the machine. There was a transformer that converted electrical current to D.C. Violano was heralded as one of the mechanical marvels of the time.

POOR	FAIR	GOOD	EXCELLENT
$9,000	$11,000	$13,000	$15,000

MILLS WIZARD FORTUNE TELLER
circa 1920

What's my fortune? Put in a penny, select one of 6 questions, and give the fortune wheel a spin. The resulting fortunes were more humorous than they were mystical.

POOR	FAIR	GOOD	EXCELLENT
$300	$400	$500	$600

FUTURE PRODUCTS FORTUNE TELLER
circa 1920

This little arcade piece was a popular stopping place for many pennies. The concept was simple: Once the penny was inserted, a three dimension phrase would appear under the glass. The accuracy of these fortunes is somewhat doubtful.

POOR	FAIR	GOOD	EXCELLENT
$200	$300	$400	$500

EXHIBIT SUPPLY CUPID'S POST OFFICE
circa 1920

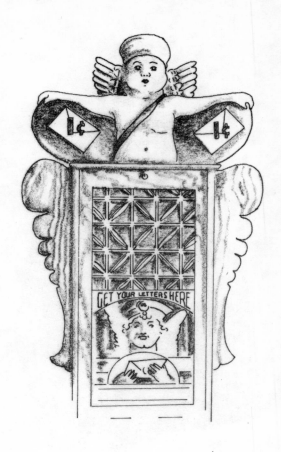

Cupid's Post Office had a coin slot for both men and women. One of 24 different letters, complete with a photograph of the sender, was delivered. The original Exhibit Supply catalog said this about the machine: "Cabinet handsomely finished in natural oak. Ornaments brightly colored papier mache and aluminum castings".

POOR	FAIR	GOOD	EXCELLENT
$1,000	$1,300	$1,600	$2,000

EXHIBIT SUPPLY GRANDFATHER'S CLOCK
circa 1927

Grip testers were popular, so the manufacturers kept coming up with variations to keep the customers interested. The Grandfather's Clock is similar to the Striking Clock; the patrons squeeze for everything they're worth, thus registering the grip on a clock dial. Then the grandfather clock sounds off the number registered.

POOR	FAIR	GOOD	EXCELLENT
$900	$1,200	$1,500	$1,800

FEY SHOOT THE DUCKS
1928

Charles Fey, of slot machine fame, was a great tinkerer. He dabbled in most types of coin-operated machines, and Shoot the Ducks was one of his ventures into the arcade field. If the trigger was pressed at the same time as the duck was passing over the appropriate contact point, the duck fell backward, giving the illusion of being hit by a bullet. The example shown, which is the only one known to exist, is under restoration.

RARE

CHESTER POLLARD PLAY THE DERBY
circa 1930

Chester Pollard games were popular in arcades. They were set up so that two people could compete against each other. In the Derby, each player turned his wheel to move his horse. The horse raced around an oval track, and the first one across the finish line was the winner.

POOR	FAIR	GOOD	EXCELLENT
$1,500	$2,000	$2,500	$3,000

EXHIBIT SUPPLY STRIKING CLOCK
circa 1930

Striking Clock was a unique concept in strength testers. The harder the squeeze, the farther the dial would register; which was typical of all strength testers. However, this one sounded the bell like a striking clock, to let everyone in the arcade know how well the participant had done.

POOR	FAIR	GOOD	EXCELLENT
$600	$800	$1,000	$1,200

JENNINGS COMET
circa 1930

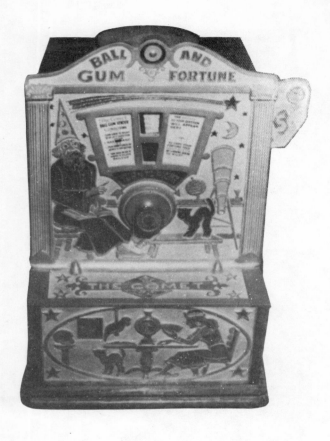

The Comet is Jenning's version of a fortune telling machine, somewhat patterned after Mills' Wizard. For 1 cent the patron asked a question, turned a dial, was given an answer, and a gumball. Notice the wizard and the fortune teller cast on the front.

POOR	FAIR	GOOD	EXCELLENT
$350	$450	$550	$700

SEEBURG CHICKEN SAM
circa 1931

Shooting galleries are arcade classics, and they are still popular with arcade patrons. Seeburg's Chicken Sam character adds a little more amusement and pizzaz to this shoot 'em up game.

POOR	FAIR	GOOD	EXCELLENT
$450	$600	$750	$900

BINGO
1931

Bingo was one of the earliest pin games. It was produced by several manufacturers, including Bingo Novelty, Gottlieb, and Field.

POOR	FAIR	GOOD	EXCELLENT
$250	$325	$400	$500

GOTTLIEB WHIZZ BANG
1932

Gottlieb was into the pinball game business very early, and his success continued for a number of years. Whizz Bang offered a brightly colored playing field, with the usual pin obstacles, and the numbered drop holes.

POOR	FAIR	GOOD	EXCELLENT
$250	$325	$400	$500

GOTTLIEB'S BIG BROADCAST
1933

Gottlieb was one of the front runners in the pinball industry. Coin-operated entertainment was becoming big business, and Gottlieb was prolific in the production of pinball games. Big Broadcast was one of his very successful early pins that played on the radio broadcasting theme.

POOR	FAIR	GOOD	EXCELLENT
$250	$325	$400	$500

RADIO STATION
circa 1933

In the early days of pinball, there were no flippers, and automatic scoring devices. The games were basically mechanical and it was a score-it-yourself system. Gambling was often associated with pinballs, and for that reason, they were later made illegal in certain states. Radio Station was a bright, inviting game that played on the radio theme which was in its heyday.

POOR	FAIR	GOOD	EXCELLENT
$250	$325	$400	$500

CUPID'S POST OFFICE
circa 1935

Cupid's Post Office is a sort of lonely heart's club for the person of small means. For 5 cents (later updated to 10 cents) the patron received a letter appropriately addressed to gents or ladies. The letters were flowery, and filled with endearments of love.

POOR	FAIR	GOOD	EXCELLENT
$600	$800	$1,000	$1,200

BABE RUTH BASEBALL
circa 1935

Babe Ruth Baseball was created by Stephen's Novelty Company of Milwaukee. The object of the game was to move the baseball player into position, and catch the ballbearing falling through the playing field. Catching all 5 balls gave the highest score possible.

POOR	FAIR	GOOD	EXCELLENT
$300	$400	$500	$600

CHALLENGER DUCKSHOOT
circa 1940

One of the simplest arcade machines ever produced was the Challenger. Put in a penny and a ballbearing dropped into mechanical gun: then try to shoot the duck. Very simply, it was a poor man's shooting gallery.

POOR	FAIR	GOOD	EXCELLENT
$100	$135	$170	$200

EXHIBIT SUPPLY NOVELTY MERCHANTMAN
circa 1942

Claw machines have always been a favorite in arcades. Put a coin in and the shovel could be directed to a prize. If the players were lucky, and they usually weren't, they would grab the prize they wanted, and direct it back over to the chute. Exhibit supply sold this model in large numbers.

POOR	FAIR	GOOD	EXCELLENT
$600	$800	$1,000	$1,200

MILLS OWL
1942

Mills was mostly devoted to the production of slot machines, but surprisingly enough they did dabble in the pinball market. The back glass of this pinball does have a definite similarity to a slot machine and this model was produced in both payout, and non-payout types.

POOR	FAIR	GOOD	EXCELLENT
$300	$400	$500	$600

MUTOSCOPE IT'S A KNOCKOUT
circa 1949

It's a Knockout is an exciting, two player action game. Players on each side of the machine maneuver a control handle. The object is to have your boxer punch the other boxer on the chin, and score a knockout.

POOR	FAIR	GOOD	EXCELLENT
$500	$650	$800	$1,000

BOOZ BAROMETER
circa 1950

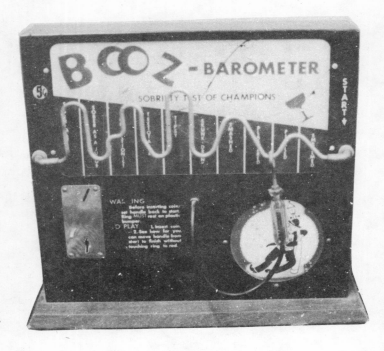

A late arcade game that has an obvious relationship to drinking establishments. The idea is to start at one end with the handle and move it to the other end without touching the ring to the rod. Failing at this test sets off lights and buzzers.

POOR	FAIR	GOOD	EXCELLENT
$75	$100	$125	$150

KEENEY'S LEAGUE LEADER
1951

Keeney was one of the smaller pinball manufacturers. In this popular game the player pushed the button to release the ball and with the other button he swung a bat. The hit balls flew into various tiers, scoring hits, runs, or outs.

POOR	FAIR	GOOD	EXCELLENT
$250	$325	$400	$500

GOTTLIEB 4 BELLES
1954

4-Belles is one of Gottlieb's early flipper games. Flipper games added a whole new dimension in pinballs. Flippers made the element of skill much more important. Notice the interesting lithography of the back glass.

POOR	FAIR	GOOD	EXCELLENT
$250	$325	$400	$500

GOTTLIEB'S GOLD STAR
1954

Gottlieb's Gold Star falls in the oak rail, early flipper game era. Kids growing up in the 1950's remember these games well. Most would rather play one of these simple games over the highly sophisticated, electronic marvels of today. Pretty girls are predominant on the back glass of this game, and they are a recurring theme on most games up through the present.

POOR	FAIR	GOOD	EXCELLENT
$200	$250	$300	$400

VENDING MACHINES

Vending machines began to become popular in America in the late 1800's. Automation was beginning to appear everywhere, and the concept of an automatic clerk was a natural. The advantages of these contraptions were obvious. The machines were always on duty, they collected no wage, and face to face contact with the public was unnecessary.

The most common product vended was gum. It came in sticks of various sizes, rectangular blobs called tabs, and brightly colored gum balls. In the early machines of the 1890's, through 1910, wrapped sticks, and tab gum seemed to be the most prevalent. One of the favorites of the time was Adams, with their famous tutti-fruitti. There were a number of others, such as Zeno, Colgans, and Mansfield, all of which have disappeared over the years. It's interesting to note that this marketing concept had a huge percentage of the country chewing gum, a habit which disgusted Europeans.

The popularity of vending gum through a machine made way for vending a number of other products through machines. Cigars were a popular product of the times. Many men felt that it was masculine, and successful-looking to smoke cigars. Various cigar vendors began to appear, incorporating their own cigar cutters. Of course, since the customer now has a cigar, he needs a match to light it with. Inevitably, match vendors were popular. It's peculiar that matches were sold in the 1890's, and they are given away in the 1980's.

Another convenience product that was vended early in the century was collar buttons. Collar buttons were used to attach detachable collars on men's shirts. Detachable collars made it possible to have a clean collar, without laundering the whole shirt. Unfortunately, collar buttons were small, and easily lost, so the collar vendor served a need.

Over the years, machine manufacturers have produced vendors for almost everything imaginable. The vendors knew that in order to sell their products, the machine and product needed to attract the customer's attention. Many of the machines were beautifully ornate, and brightly painted. Some had animated figures that entertained the customer while he or she received the product. Many had mechanisms that were fascinating to watch. The result, of course, were some incredibly beautiful machines that are unique works of art. Unfortunately over the years, the ornate machines have given way to more streamlined, no-nonsense vendors that aren't much fun. However, there seems to be no slowdown of the American public's willingness to buy products from a machine.

VENDING MACHINES INDEX

Illustration page number — **Machine name** — **Price range**

231	Zeno Gum c. 1895	$325-$650
232	Adam's Tutti Fruitti c. 1900	$500-$1,000
233	Price Collar Button Vendor 1901	$425-$950
234	Mansfield Automatic Clerk 1901	$250-$500
235	National Vending Colgan's Gum c. 1902	$900-$1,800
236	Happy Jap 1902	$750-$1,500
237	Doremus Cigar Vendor c. 1907	$1,000-$2,000
238	International Vending Match Machine c. 1910	$600-$1,250
239	Postage Vendor c. 1910	$175-$350
240	Zeno Gum c. 1910	$175-$350
241	Advance 1 cent Matches 1915	$225-$450
242	Advance Gumball c. 1919	$125-$250
243	The Leebold c. 1920	$600-$1,200
244	Diamond Book Matches c. 1920	$225-$450
245	C.E. Leebold c. 1920	$80-$165
246	Gaylord Manufacturing Scoopy Gum Vendor c. 1920	$250-$500
247	Climax 1920	$250-$550
248	Columbus "B" c. 1920	$75-$150
249	Columbus "B" Slug Ejector c. 1920	$100-$200
250	Columbus Model 21 c. 1920	$100-$200
251	Columbus "A" c. 1920	$125-$250
252	Tom Thumb c. 1920	$50-$100
253	Advance Gumball 1923	$85-$175
254	Columbus Bi-More c. 1925	$200-$400
255	Log Cabin Duplex Vendor c. 1927	$60-$125
256	The Old Mill c. 1928	$375-$750
257	Pulver c. 1930	$150-$300
258	Hawkeye c. 1930	$50-$125
259	The Vendor c. 1930	$50-$110

Illustration page number	Machine name	Price range
260	Master 1 cent c. 1930	$75-$150
261	Master 1-5 cent c. 1930	$100-$225
262	Master Fantail 1-5 cent c. 1930	$200-$500
263	Simmons Model A c. 1930	$75-$150
264	Oronite Lighter Fluid Dispenser c. 1930	$100-$250
265	Exhibit Supply Company card vendor c. 1930	$75-$150
266	Northwestern Merchandiser 31 1931	$75-$150
267	Smilin' Sam 1931	$375-$750
268	Ad-lee EZ c. 1932	$300-$600
269	Northwestern 33 Jr. 1933	$150-$300
270	Northwestern 33, 1933	$60-$125
271	Northwestern 33, 1933	$50-$100
272	Abbey Gum Vendor c. 1935	$30-$75
273	Columbus "B" Bulk Vendor c. 1935	$65-$125
274	Regal Gumball Vendor c. 1935	$30-$75
275	Star Gum Vendor c. 1935	$60-$125
276	Tray Style Vendor c. 1940	$60-$125
277	Victor Baby Grand c. 1940	$20-$50
278	Victor Model V c. 1940	$30-$75
279	Adams Gum Vendor c. 1940	$60-$125
280	Oak Acorn c. 1950	$30-$60
281	Regal Hot Nut Vendor c. 1950	$30-$75
282	Ohio Book Matches Dispenser c. 1950	$20-$50
283	Northwestern Vendor c. 1950	$25-$50
284	Premiere Card Vendor c. 1950	$30-$75
285	Ford Chrome c. 1950	$20-$40

ZENO
circa 1895

Zeno machines were always a little unusual. For example this machine offered the following flavors; Pineapple, yucca, licorice, peppermint, lemonade, vanilla cream, and pepsin. There was no choice however; once the penny was inserted a clock mechanism pushed the gum out of the column and it fell down a chute into the hands of the customer.

POOR	FAIR	GOOD	EXCELLENT
$325	$425	$525	$650

ADAMS' TUTTI-FRUITTI
circa 1900

That's right; we have the Adams' Gum Company to thank for the expression tutti-fruitti. Adams' had other flavors, but tutti-fruitti was his most popular, and the one he advertised the most. This is one of Adams' earliest machines. Note the beautiful quarter sawn oak case.

POOR	FAIR	GOOD	EXCELLENT
$500	$650	$800	$1,000

PRICE COLLAR BUTTON VENDOR
1901

Little columns of collar buttons displayed the wares for this machine. Much like a modern vendor, the customer selected the style he wanted by moving the selector wheel. For those people out there who don't have the faintest idea what a collar button is, they were used to attach removable collars to gentlemen's shirts.

POOR	FAIR	GOOD	EXCELLENT
$425	$575	$700	$950

MANSFIELD AUTOMATIC CLERK
1901

The Automatic Clerk is a handsome machine with an etched front It was unique in that the clock wound mechanism was completely visible to the customer. A penny would set the escalator in motion, flipping the gum over the Top of the column, and ringing a bell.

POOR	FAIR	GOOD	EXCELLENT
$250	$325	$400	$500

NATIONAL VENDING COLGAN'S GUM
circa 1902

Taffy tolu had to be an interesting gum flavor. These rare machines gave 1 stick for a penny. As an extra inducement, 2 sticks were given for every 5th penny.

POOR	FAIR	GOOD	EXCELLENT
$900	$1,200	$1,500	$1,800

HAPPY JAP
1902

For 1 cent the happy Jap would shove a gum through his mouth to the awaiting customer. This was accomplished by a small clockwork motor mechanism. The machine was made of cast iron, and painted in vivid colors.

POOR	FAIR	GOOD	EXCELLENT
$750	$1,000	$1,250	$1,500

DOREMUS CIGAR VENDOR
circa 1907

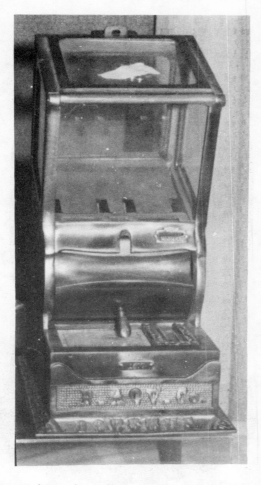

A very simple machine that allowed the cigar to fall into the tray because of its slanted top. A coin released the lever and allowed the customer to receive his purchase.

	with cigar cutter		
POOR	**FAIR**	**GOOD**	**EXCELLENT**
$1,000	$1,300	$1,600	$2,000

	without cigar cutter		
POOR	**FAIR**	**GOOD**	**EXCELLENT**
$750	$1,000	$1,250	$1,500

INTERNATION VENDING MATCH MACHINE
circa 1910

The cylinder revolved within the glass dome and dropped a box of matches into the shoot. Part of the fun in buying these matches was to watch the machine work.

POOR	FAIR	GOOD	EXCELLENT
$600	$800	$1,000	$1,250

POSTAGE VENDOR
circa 1910

Postage stamps were a natural for vending machines; everyone hates to run to the post office just for stamps. This early model has a cast iron base and top. Notice the great use of beveled glass for displaying the stamps and mechanism.

POOR	FAIR	GOOD	EXCELLENT
$175	$225	$275	$350

ZENO GUM
circa 1910

Decorated in bright yellow porcelain, the Zeno was an attention getting machine. After a penny was dropped into the slot, a clockwork mechanism pushed the gum over the top of the column, and into the product tray. A small viewing window reassured the customer that there was gum in the machine.

POOR	FAIR	GOOD	EXCELLENT
$175	$225	$275	$350

ADVANCE 1¢ MATCHES
1915

Advance displayed their matches with a beautiful glass dome. The base was cast iron with ornate litte feet. It's amazing how something as simple as a match vendor could be made to look so classy.

POOR	FAIR	GOOD	EXCELLENT
$225	$300	$375	$450

ADVANCE GUMBALL
circa 1919

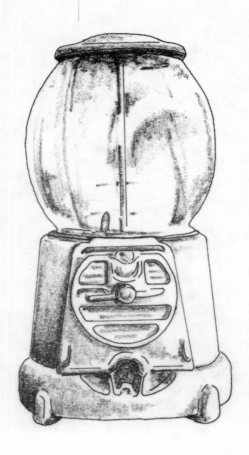

This early model Advance gumball is noteworthy because of its unusual football type globe. The more common globe was the round type which was produced in much larger numbers. Metal parts in this machine were primarily cast iron.

POOR	FAIR	GOOD	EXCELLENT
$125	$165	$200	$250

THE LEEBOLD
circa 1920

This Leebold looks older than it is because of its beautiful Victorian styling which had a revival in the 20's. One telltale sign of its age is the fact that it is made of cast aluminum. Unfortunately they are quite rare and few are to be seen around. What else can be said, it's a classic.

POOR	FAIR	GOOD	EXCELLENT
$600	$800	$1,000	$1,200

DIAMOND BOOK MATCHES
circa 1920

One penny and a turn of the dial gave the customer 1 to 4 packs of book matches. The merchant had the option of setting the vendor to deliver 1 to 4 books. Notice the window in the upper left corner that was changeable to announce how many books were being vended.

POOR	FAIR	GOOD	EXCELLENT
$225	$300	$375	$450

C.E. LEEBOLD
1920's

Leebold was noted for it's huge globes that held a lot of product. This made the operator happy, since it meant fewer refills. The base was made of cast aluminum. Note that the marquee is missing on this model.

POOR	FAIR	GOOD	EXCELLENT
$80	$110	$140	$165

GAYLORD MANUFACTURING
SCOOPY GUM VENDOR
circa 1920

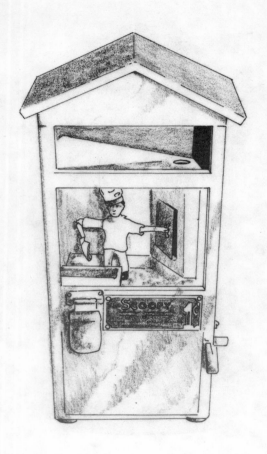

The Scoopy Gum Vendor delivers entertainment for the price of a gumball. A figural baker automated by a clock mechanism opens the oven door, and scoops a gumball into the shoot.

POOR	FAIR	GOOD	EXCELLENT
$250	$335	$420	$500

CLIMAX
1920

This cast iron based Climax has very nice lines. Climax is noted for its unique machines. This model shows a transition from the earlier, more ornate machines, to the more streamlined styling. This was the last Climax produced.

POOR	FAIR	GOOD	EXCELLENT
$250	$350	$450	$550

COLUMBUS "B"
circa 1920

This Columbus bulk vendor features the round globe, and the bell bottom base. One of the most common of the Columbus machines, because they were produced in large numbers. Note the original decal, and the embossed Columbus star on the globe. Early models were cast iron, and later models were cast aluminum.

POOR	FAIR	GOOD	EXCELLENT
$75	$100	$125	$150

COLUMBUS "B" BULK VENDOR WITH SLUG EJECTOR
circa 1920

The most unusual feature of this Columbus model "B" is its add-on slug ejector. These could be specially ordered from the company. Slug ejectors were seldom used, and it can be assumed that this machine probably came from a location with undesirable clientele.

POOR	FAIR	GOOD	EXCELLENT
$100	$135	$170	$200

COLUMBUS MODEL 21
circa 1920

This Columbus features a somewhat different base design than most of the other Columbus machines. The model shown includes the Columbus star hexagonal globe, and original barrel locks.

POOR	FAIR	GOOD	EXCELLENT
$100	$135	$170	$200

COLUMBUS "A"
circa 1920

This Columbus hourglass cast iron is definitely recognized as a classic. Fortunately they were produced in large enough numbers that prices are not exorbitant. Note the embossed Columbus star on the globe and the Columbus barrel locks.

POOR	FAIR	GOOD	EXCELLENT
$125	$165	$200	$250

TOM THUMB
circa 1920's

Miniature gumball machines would never make it today because of theft problem. However, in the good old days, this little guy was great for the location with limited counter space.

POOR	FAIR	GOOD	EXCELLENT
$50	$65	$80	$100

ADVANCE GUMBALL
1923

Advance manufacturing of Chicago, Illinois had reasonable success with their simple reliable machines. Metal base parts were tin and the front plate was nickel over brass. They came painted in a variety of bright colors. However, red was the most popular, as was the case with many of the other manufacturers.

POOR	FAIR	GOOD	EXCELLENT
$85	$115	$145	$175

BIMORE COLUMBUS DOUBLE VENDOR
circa 1925

Double vendors started to get popular in the 1920's. The idea of course, was to give the customer more choices. This Columbus was porcelain over cast iron. It vended ball gum on one side, and any bulk product, such as peanuts, on the other.

POOR	FAIR	GOOD	EXCELLENT
$200	$275	$325	$400

LOG CABIN DUPLEX VENDOR
circa 1927

Giving the customer two choices brought more money to the operators. Log Cabin was made of aluminum, and it was polished to catch a potential buyer's attention. Notice the very modern lines that was an unusual style for the 1920's.

POOR	FAIR	GOOD	EXCELLENT
$60	$80	$100	$125

THE OLD MILL
circa 1928

International Mutoscope, famous for their flip card machines, also produced the Old Mill. For a mere penny, the customer got a scoop full of candy, gum, or whatever. Because of the random nature of the scoop, the customer experienced a mild sensation of gambling. The unusual graphics are especially noteworthy on this machine.

POOR	FAIR	GOOD	EXCELLENT
$375	$500	$625	$750

PULVER
circa 1930

Pulver vendors were produced for many years, and it's easy to understand why. Kids as well as adults love to put their money in and watch them go. After inserting a penny, a clock movement set a figural character into motion, and a stick of gum would drop into the tray.

POOR	FAIR	GOOD	EXCELLENT
$150	$200	$250	$300

HAWKEYE
circa 1930

Hawkeye featured a 6-sided cast aluminum base, usually painted in bright red. The gate, and the coin mechanisms were polished aluminum, which helped to dress up the machine.

POOR	FAIR	GOOD	EXCELLENT
$50	$75	$100	$125

THE VENDOR
circa 1930

The cast iron base on this model was standard, and jobbers changed the top around to fit their needs. It's interesting how the variety of globes could change the entire look of a machine. Notice the primitive look of the Mason jar shape globe shown.

POOR	FAIR	GOOD	EXCELLENT
$50	$70	$90	$110

MASTER 1¢
1930

Norris manufacturing made a name for itself with the Master gumball and bulk vendors of the style pictured. The top and base were finished in a bright porcelain, usually black, green, or white. The square globe design was a plus in that if it was broken it was easily replaced with a piece of standard glass.

POOR	FAIR	GOOD	EXCELLENT
$75	$100	$125	$150

MASTER 1¢/5¢
circa 1930

The Master 1-5 cent is not to be confused with the more valuable Master 1-5 cent fantail. With this machine the penny and nickel were both inserted in the same slot. A penny received one measure, and a nickel received a little more than 5 times that amount. The top and bottom of the machine were made of porcelain for extra durability.

POOR	FAIR	GOOD	EXCELLENT
$100	$150	$175	$225

MASTER FANTAIL 1¢/5¢
circa 1930

Master Fantail is a classic that is sought after by collectors. A customer could deposit a penny in one slot, or a nickel in another, and receive varied amounts of the product. The finish was porcelain, and it came in several different colors. Note that the patent date on all Master machines says 1923; but they weren't actually distributed until 1930.

POOR	FAIR	GOOD	EXCELLENT
$200	$300	$400	$500

SIMMONS MODEL A
circa 1930

This Simmons instructed the customer in 1-2-3 sequence on the front casting. Number 1, insert coin, number 2, push lever, number 3, turn knob. If the buyer wasn't tired after all that, he could open the goods door, and receive the merchandise. Notice the geometric designs in the globe, and the porcelain finish.

POOR	FAIR	GOOD	EXCELLENT
$75	$100	$125	$150

ORONITE LIGHTER FLUID DISPENSER
circa 1930

Notice the resemblance to an early gas pump? For 1 cent the customer was treated to a fillup for his lighter. Why don't we see these vendors anymore? One reason could be that their diminutive size opens up all kinds of theft possibilities.

POOR	FAIR	GOOD	EXCELLENT
$100	$150	$200	$250

EXHIBIT SUPPLY COMPANY - CARD VENDOR
circa 1930

This card machine was set up to vend novelty cards. The particular one shown vends a card showing your future husband or wife, and children. If that idea doesn't appeal to the customer, there's a lonely hearts club card.

POOR	FAIR	GOOD	EXCELLENT
$75	$100	$125	$150

NORTHWESTERN MERCHANDISER 31
1931

Northwestern put porcelain on most of their machines. The reason of course was to have a durable and attractive finish. This merchandiser is cast iron, with a blue and yellow porcelain finish. Note the original Northwestern decal.

POOR	FAIR	GOOD	EXCELLENT
$75	$100	$125	$150

SMILIN' SAM
1931 (1975)

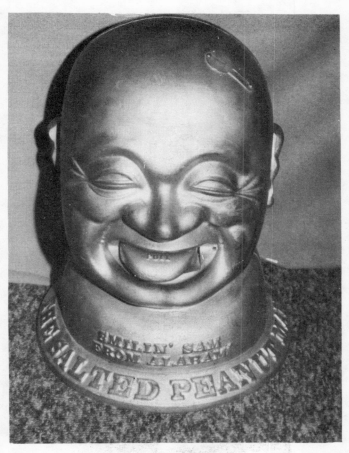

Smilin' Sam was first brought out in 1931, and later reproduced in 1975. Insert your coin, pull his tongue, and he drops a measure of peanuts into your hand. This machine is very similar to the Happy Jap in concept, so it's very likely that it was a case of one company borrowing another's idea.

1931

POOR	FAIR	GOOD	EXCELLENT
$375	$500	$625	$750

1975

POOR	FAIR	GOOD	EXCELLENT
$150	$200	$250	$300

AD-LEE E Z
circa 1932

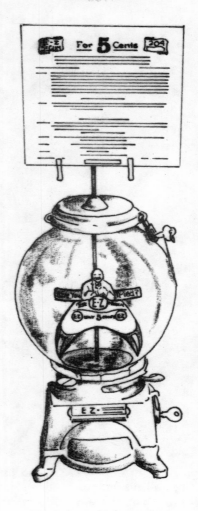

Is it a gambling machine, or a gumball machine? It's two machines in one. In a very unique concept, the Adlee company supplied hollow gumballs with a slip of paper inside. The slip of paper had a horse's name printed on it. If the name corresponded to the award card on the machine's marquee, the customer received a trade award. Of course, the gumball was a fairly expensive 5 cents.

POOR	FAIR	GOOD	EXCELLENT
$300	$400	$500	$600

NORTHWESTERN 33 JR
1933

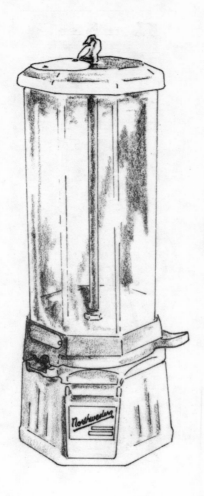

Almost all of the manufacturers produced baby machines, or juniors. These machines are cute, and they are often sought after by collectors. This model had a hexagon globe, and came with porcelain over cast iron. They came in all colors, but light green was the most common.

POOR	FAIR	GOOD	EXCELLENT
$150	$200	$250	$300

NORTHWESTERN "33"
1933

Porcelain finishes were used on all of the early Northwestern machines. This finish was durable and held up to the hard knocks of being on location. Many examples of the Northwesterns have survived with their original porcelain, thus giving the modern-day collector a shot at having a machine with an all-original finish.

POOR	FAIR	GOOD	EXCELLENT
$60	$80	$100	$125

NORTHWESTERN 33
1933

Northwestern used a logical approach for model numbers since they were simply the year of production. This model was porcelain (usually bright red) over cast iron. Note the original decal inviting the customer to "try some".

POOR	FAIR	GOOD	EXCELLENT
$50	$65	$80	$100

ABBEY GUM VENDOR
circa 1935

The Abbey Gum Vendor is a nice-looking little machine, especially when its aluminum finish is buffed up like chrome. The customer slid his money into the cash tray and the machine delivered the product to the plate-like tray. The Abbey is very small in size, and it was designed for limited space locations.

POOR	FAIR	GOOD	EXCELLENT
$30	$45	$60	$75

COLUMBUS "B" BULK VENDOR
circa 1935

Columbus manufactured a huge variety of gumball and bulk vending machines. The one pictured is a Bell Bottom bulk vendor with a hex globe. Note the original barrel locks; they are important to the value of the machine. Original decals are also a big plus feature.

POOR	FAIR	GOOD	EXCELLENT
$65	$85	$100	$125

REGAL GUMBALL VENDOR
circa 1935

Modular design and a teardrop globe identify the Regal vendor. This model was aluminum, as was the case with most gumball machines produced in the late-1930's.

POOR	FAIR	GOOD	EXCELLENT
$30	$45	$60	$75

STAR GUM VENDOR
circa 1935

The Star Gum Vendor has an interesting shape, with its triangular hexagonal globe. Star Quality Gum was advertised to have flavor all through. Whether or not this was a false advertising claim is hard to say. It probably was, however, since false advertising claims were common in early sales promotions.

POOR	FAIR	GOOD	EXCELLENT
$60	$80	$100	$125

TRAY STYLE VENDOR
circa 1940

Several manufacturers came out with tray style vendors for the customer that was clumsy. This way, the product could be scooped up from the immaculately clean tray, if spilled. The machine was made of aluminum, and painted in two tones.

POOR	FAIR	GOOD	EXCELLENT
$60	$80	$100	$125

VICTOR BABY GRAND
circa 1940

The Baby Grand was produced in a variety of models, including card vendor, jawbreaker, novelty, bulk, and gumball models. The sides were varnished oak, and that added a nice touch. Unfortunately the viewing portions are made of plastic, instead of glass.

POOR
$20

FAIR
$30

GOOD
$40

EXCELLENT
$50

VICTOR MODEL V
circa 1940

Victor represents the no nonsense, boxey shapes of the 1940's. These were good working, inexpensive machines, and they were produced in large numbers. The original decal on the machine pictured adds to the value of the machine.

POOR	FAIR	GOOD	EXCELLENT
$30	$45	$60	$75

ADAM'S GUM VENDOR
circa 1945

Adam's gum Company was an old hand at the business of vending gum. This is one of their innovations for selling gum. A multiple row concept much like cigarettes, gave the customer a great deal of choice, with 6 different selections.

POOR	FAIR	GOOD	EXCELLENT
$60	$80	$100	$125

OAK ACORN
circa 1950

Almost any kid who has ever purchased a gumball, or bulk-vend candy will recognize the acorn symbol. Oak Mfg. was king of the gumball vendors in the 40's and 50's and beyond. Their symbol can still be seen on gumball machines across the country.

POOR	FAIR	GOOD	EXCELLENT
$30	$40	$50	$60

REGAL HOT NUT VENDOR
circa 1950

The special attraction with this vendor was its function of vending hot nuts. A light bulb served to keep the nuts warm, and attract the attention of passersby. This is an interesting machine that's a little different from the run of the mill vendors. A glass light cover indicated the machine was made in the 1930's, and a plastic cover indicates the 1950's.

Plastic

POOR	FAIR	GOOD	EXCELLENT
$30	$45	$60	$75

Glass

POOR	FAIR	GOOD	EXCELLENT
$75	$100	$125	$150

OHIO BOOK MATCHES DISPENSER
circa 1950

This match book vendor was a simple little device that was prevalent in restaurants, smoke shops, pool rooms, etc. The ornate castings and glass viewing areas that were part of earlier match vendors, are conspicuously absent in this vendor.

POOR	FAIR	GOOD	EXCELLENT
$20	$30	$40	$50

NORTHWESTERN VENDOR
circa 1950's

Like most of the other gumball manufacturers, Northwestern updated their machines in later years. Notice the very modular design with the shiny chrome finish. The viewing area also made use of plastic, instead of glass.

POOR	FAIR	GOOD	EXCELLENT
$25	$30	$40	$50

PREMIERE CARD VENDOR
circa 1950

The Premiere vended the customer both a ball gum, and a personality card. Personality cards were big in the 1950's, and many kids were hooked on collecting sets. Interestingly enough, some of these sets of cards are beginning to have value today.

POOR	FAIR	GOOD	EXCELLENT
$30	$45	$60	$75

FORD "CHROME"
circa 1950

The Ford Chrome is familiar to almost everyone who has ever purchased a gumball. Many of these machines are still on location, however, most have been converted to plastic globes. Because of their great numbers and late vintage, they are very available and fairly inexpensive.

POOR	FAIR	GOOD	EXCELLENT
$20	$30	$35	$40

BIBLIOGRAPHY
AND FURTHER SUGGESTED READING

BOOKS

Drop Coin Here by Ken Rubin, Crown Publishing, One Park Avenue, New York, N.Y. 10016

Encyclopedia of Automatic Musical Instruments by Q. David Bowers, Vestal Press, Box 97, Vestal, New York, 13850

An Illustrated Price Guide to the 100 Most Collectible Trade Stimulators by Richard M. Bueschel, Coin Slot Books, Box 612, Whearidge, Colorado 80033

An Illustrated Price Guide to the 100 Most Collectible Slot Machines, by Richard M. Bueschel, Coin Slot Books, Box 612, Wheatridge, Colorado 80033

An Illustrated Price Guide 100 Most Collectible Slot Machines - Volume 2, by Richard M. Bueschel, Coin Slot Books, Box 612, Wheatridge, Colorado 80033

Jukebox Saturday Night, by J. Krivine, Chartwell Books, Inc., 110 Enterprise Ave., Secaucaus, New Jersey 07094

The Official Loose Change Blue Book of Antique Slot Machines, by Stan and Betty Wilker, The Mead Company, 21176 South Alameda St., Long Beach, California 90810

The Official Loose Change Red Book of Antique Trade Stimulators and Counter Games, by Stan and Betty Wilker, The Mead Company, 21176 South Alameda St., Long Beach, California 90810

Owner's Pictorial Guide for the Care and Understanding of the Mills Bell Slot Machine, by Robert N. Geddes and Daniel R. Mead, The Mead Company, 21176 South Alameda St., Long Beach, California 90810

Pinball Portfolio, by Harry McKeown, Chartwell Books, 110 Enterprise Avenue, Secaucus, New Jersey, 07094

Slot Machines on Parade, by Robert N. Geddes and Daniel R. Mead, The Mead Company 21176 South Alameda St., Long Beach, California 90810

Slot Machines, A Pictorial Review, by David G. Christensen, The Vestal Press, Box 97, Vestal, New York, 13850

Tilt The Pinball Book, Home Maintenance, by Candace Ford Tolbert and Jim Alan Tolbert, Creative Arts Book Company, 833 Bancroft Way, Berkeley, Ca. 94710

PERIODICALS

"The Antique Trader", Weekly tabloid, misc. antique classifieds, The Antique Trader, Box 1050, Dubuque, Iowa

"The Coin Slot", monthly magazine, coin operated antiques, articles and classified ads, Bill Harris, P.O. Box 612, Wheatridge, Colorado 80033

"The Jukebox Trader", monthly newsletter, jukebox ads and articles, Jukebox Trader Box 1081, Des Moines, Iowa 50311

"Loose Change", monthly magazine, coin operated antiques, articles and classified ads, Mead Company, 21176 South Alameda St., Long Beach, California, 90810

CATALOGUES

Mechanical Music Center, Inc., 25 Kings Hwy. North, Box 88, Darien Conn., 06820; very nice illustrated catalogue of musical instruments (some coin-ops)

Coin Slot Library list; books, old catalog reprints, posters, service manuals on antique coin-operated machines. The Coin Slot, P.O. Box 612, Wheatridge, Colo. 80033

East Coast Casino Antiques Catalogue; very nice, illustrated booklet of primarily gambling paraphernalia with some coin-operated gambling devices; East Coast Casino Antiques, 98 Main St., Fishkill, New York 12524

Jukebox Junction; illustrated catalogue of reproduction jukebox parts and literature with related items; Jukebox Junction, Box 1081, Des Moines, Iowa

Misc. lists of books, reprinted catalogues, posters, and machines for sale; Mr. Russell, 2404 West 111th St., Chicago, Illinois 60655

list of reprints: reel strips, paycards, decals, instruction sheets, and mint wrappers for slots and trade stimulators. D. B. Evans, 7999 Keller Rd., Cincinnati, Ohio 45243

list of reprints: reel strips, paycards, decals, instruction sheets, and Vestal Press Catalogue; illustrated catalogue of books on antiques, with some on coin-operated antiques; The Vestal Press Box 97 Vestal, New York, 13850

list of reprints: reelstrips and paycards for trade stimulators. Slot Dynasty, 23 Palmdale Ave., Daly City, Ca. 94015

MISC.

Old catalogue reprints are a valuable source for research on antique machines. Mr. Russell, and the Coin Slot library, mentioned in the catalogues list both carry reprints. Mead Publication and various other sources also carry reprints.

Back issues of Billboard magazine and the Coin Slot Journal are also valuable resource materials. Libraries in larger cities sometimes have this information on microfilm.